SPECIAL PAPERS IN PALAEONTOLOGY NO. 36

CAMPANIAN AND MAASTRICHTIAN AMMONITES FROM NORTHERN AQUITAINE, FRANCE

BY

W. J. KENNEDY

with 23 plates and 43 text-figures

THE PALAEONTOLOGICAL ASSOCIATION
LONDON
DECEMBER 1986

Printed in Great Britain

CONTENTS

ABSTRACT. The ammonite fauna of the Campanian Stage in the Aquitaine Basin is revised on the basis of the classic collections of Coquand, Arnaud, and others. Thirty-eight species are described, and referred where possible to the standard fourfold division of the stage, with zones of *Placenticeras bidorsatum* (oldest), *Menabites (Delawarella) delawarensis*, *Hoplitoplacenticeras (Hoplitoplacenticeras) marroti*, and *Nostoceras (Bostrychoceras) polyplocum*. A limited Maastrichtian assemblage, dated on microfossil criteria, is recognized from the highest part of the section with *Sphenodiscus ubaghsi* de Grossouvre, 1894, *N. (Nostoceras) hyatti* Stephenson, 1941, and *Baculites leopoliensis* Nowak, 1908 occurring above the rudist-bearing Dordonian *sensu stricto* of Coquand. A single specimen of *S. ubaghsi* is from below this level but there is no unequivocal ammonite evidence to date the Dordonian *sensu stricto* as Campanian *versus* Maastrichtian.

INTRODUCTION

IN recent publications I have, with colleagues, revised the ammonite faunas of the Upper Cretaceous stages in their type areas. This contribution completes this study, begun some fifteen years ago. I here describe the ammonites of the Campanian Stage, the longest division of the Upper Cretaceous, in its type region of northern Aquitaine (text-fig. 1). Also included are the few ammonites known from the succeeding Cretaceous sediments, which are dated as early Maastrichtian age on microfossil criteria.

The revision of faunas from the Cenomanian to Coniacian stages revealed a wealth of new material and of stratigraphic information; the Santonian less so. With the Campanian, my own work in Aquitaine has revealed only a handful of ammonite fragments, and this revision relies on the old collections, notably those of Arnaud, in turn the basis of much of de Grossouvre's (1894) great work on the ammonites of the 'Craie Supérieur' of France. Many of these specimens came from quarries and roadcuts now long vanished, or from the deep trenches and tunnels of the railway system that traverses the region, all now equally overgrown. I thus rely heavily on the results of my predecessors, and this stratigraphic introduction is correspondingly brief. This is a great disappointment, for recent estimates give the duration of the stage as 12·5 Ma (Hallam *et al.* 1985). As Hancock and I suggested some time ago (1981, p. 542) this would lead to an expected twenty to twenty-five ammonite zones, yet only three to four have been generally recognized in Western Europe and only four, in essence those first used by de Grossouvre (1901) and Haug (1907–1911), are applicable to the Aquitaine succession.

SUBDIVISION OF THE SENONIAN IN AQUITAINE

D'Orbigny (1852, pp. 666–691) cited a number of localities in Aquitaine when he discussed the Senonian Stage, including Royan, Cognac, and Saintes. Subdivisions of the sequence were recognized by d'Archiac (1836, 1843, 1851, 1857) and Coquand (1856 and later references). Coquand (1857*a–c*, 1858) divided d'Orbigny's Senonian into four successive stages, Coniacien, Santonien, Campanien, and Dordonien. These names replaced the number system introduced by Coquand in 1856 (Table 1), and were defined as follows (1857*b*, p. 748):

B. CRAIE SUPÉRIEURE.

Premier étage.—Coniacien.

A. Sables et grès de Richemont (*Ostrea auricularis*, Coquand).

B. Calcaire chloriteux (*Ostrea auricularis*, Coq., *Ammonites polyopsis*, Dujard., *A. Bourgeoisi*, d'Orb. *Terebratula Arnaudi*, Coq., *T. coniacensis*, Coq., *Rhynchonella Baugasi*, d'Orb., *Sphaerulites Coquandi*, Bayle *Micraster brevis*, Agass.).

Cet étage correspond au sixième horizon de rudistes.

Deuxième étage.—Santonien.

Craie tendre avec silex (*Pleurotomaria santonesa*, d'Orb., *Janira Truellei*, d'Orb., *Spondylus hippuritorum*, d'Orb., *Rhynchonella vespertilio*, d'Orb., *Hemiaster stella*, Desor, *Salenia geometrica*, Agass.).

Troisième étage.—Campanien.

Craie, tendre (*Ostrea vesicularis*, Lam., *O. larva*, Lam., *Sphaerulites Hoeninghausi*, Des Moul., *Ananchytes ovata*, Lam.).

Cet étage correspond au septième étage de rudistes.

Quatrième étage.—Dordonien.

Craie avec *Hippurites radiosus*, Des Moul., *Sphaerulites cylindraceus*, Des Moul., *Radiolites Jouanneti*, d'Orb.

Cet étage correspond au huitième horizon de rudistes.

TABLE 1. Correlation of previous subdivisions of the Campanian–Dordonian in northern Aquitaine, based on the work of Séronie-Vivien (1972, table 1)

COQUAND 1856 1re nomenclature	COQUAND 1857 3e nomenclature	ARNAUD 1877	
3e ETAGE 7e horizon de Rudistes Calcaire jaune à Sphaerulites cylindraceus, Radiolites Jouanneti, et Hippurites radiosus	4e ETAGE: DORDONIEN 9e horizon de Rudistes Ostrea lameraciana, Hippurites radiosus, Radiolites Jouanneti, R. Bournoni, Sphaerulites Toucasi, S. cylindraceus	S	Grès supérieur dolomitique et poudingue marneux à Rudistes Hippurites Lamarcki, Psammechinus
		R	Calcaires jaunes supérieurs. R^{b} Calc. dolomitiques. R^{a} Calc. arénacés Radiolites acuticostatus, Hemipneustes radiatus, Hemiaster prunella
		Q	Calcaire glauconieux Orbitolites media, Radiolites crateriformis, Rhynchopygus Marmini
2e ETAGE 6e horizon de Rudistes Craie tendre à O. vesicularis, O. larva, Sphaerulites Hoeninghausi, Radiolites crateriformis, Ananchytes ovata (Grande Champagne)	3e ETAGE: CAMPANIEN 8e horizon de Rudistes O. vesicularis, O. larva, Ananchytes ovata, Baculites Faujasi, Hemiaster prunella, Conoclypeus Leskei, Sph. Hoeninghausi, Sph. alatus, Radiolites royanus, R. fissicostatus	P	Calcaires marneux à silex: Calcaire supérieur blanc ou glauconieux Ananchytes ovata Calcaire moyen marneux ou arénacé Belemnitella quadrata Calcaire inférieur marneux hydraulique Hippurites Arnaudi, Rhynchonella globata
1er ETAGE 5e horizon de Rudistes 3e sous-étage Craie tendre, à silex (Petite Champagne)	2e ETAGE: SANTONIEN 7e horizon de Rudistes Micraster brevis, Spondylus truncatus, Rhynchonella vespertilio, Radiolites Arnaudi	N'	Marnes et grès santoniens supérieurs: N^{c} Calcaire glauconieux à Conoclypeus ovum N^{b} Banc à Ostrea acutirostris N^{a} Grès à Sphaerulites Hoeninghausi
		N	Marnes à Ostrea vesicularis et Ostrea proboscidea
		M	Marnes et grès santoniens inférieurs Botriopygus Toucasanus

Localities in Coquand (1856) are, strictly speaking, the stratotypes of these stages. For the Santonien, Coquand cites under Troisième sous-étage: 'les bords de la Charente jusqu' aux côteaux qui, à partir de Gimeux, Genté, Segonzac, dessinent un bourrelet saillant parallèle aux dernières rides de la craie inférieur. Ce plaine, qui prend le nom de *Petit-Champagne*... Une excellent étude de se système peut se faire dans les environs du château de Malberchie' (1856, pp. 86, 87). For the Campanien, Coquand cites under Deuxieme étage: 'la Grande-Champagne...escarpements qui domine le cours de la Dronne, à l'extrémité méridionale du départment, les environs d'Aubeterre...le village d'Archiac... Les falaises qui s'entendent depuis Mortagne jusqu' a Royan' (1856, pp. 88–92). For the Dordonien,

TABLE 1

ARNAUD 1883				DE GROSSOUVRE 1901			ZONES A:	AQUITAINE	HAUG 1911	
DANIEN Desor	DORDONIEN	S	Sables, grès et poudingues Radiolites ingens, Sphaerulites Toucasi, etc.	SENONIEN	CAMPANIEN	supérieur	Pachydiscus neubergicus	Couches à Thécidées de Maurens Silex à Faujasia de la Dordogne Bancs à Hippurites radiosus de Lamérac et du Maine Blanc Couches à Orbitoïdes Craie de Royan	MAESTRICHTIEN	Zone à Pachydiscus neubergicus
		R	Calcaires jaunes Hemipneustes, Faujasia Faujasi, etc.							
		Q	Calcaires glauconieux Orbitolites media, Conoclypeus Leskei, etc.							
	CAMPANIEN	P[3]	Calcaire blanc gélif Micraster glyphus, etc.				Hoplites vari	Marnes glauconieuses à Temnocidaris Baylei, Terebr. santon. Calcaire marneux		Zone à Bostrychoceras polyplocum
									CAMPANIEN	Zone à Hoplites vari
		P[2]	Calcaire blanc légèrement grisâtre Belemnitella quadrata			inférieur	Mortoniceras delawarense	Craie à Ammonites delawarensis et Scaphites hippocrepis		Zone à Mortoniceras delawarense
		P[1]	Calcaire hydraulique blanc ou bleuâtre, à silex Scaphites binodosus, Baculites anceps, etc.				Placenticeras bidorsatum	Calcaire marneux à Micraster regularis		Zone à Placenticeras bidorsatum
	SANTONIEN	N	Grès ou calcaire marneux, avec ou sans silex, et géodes de quartz Ammonites polyopsis, etc.		CORBIERIEN	SANTONIEN	Placenticeras syrtale	Calcaire sableux à Rudistes Marnes à Ostracées	SANTONIEN	Zone à Placenticeras syrtale
		N	Calcaire marneux avec silex, passant à des marnes grises ou verdâtres, pétries d'Ostracées O. vesicularis, etc.							
		M	Calcaire marneux ou noduleux grès calcarifère Ammonites polyopsis, etc.				Mortoniceras texanum	Banc à Botriopygus Couches à Ammonites texanus		Zone à Texanites texanum
		M	Calcaire marneux arénacé Ammonites coniacencis, etc.							

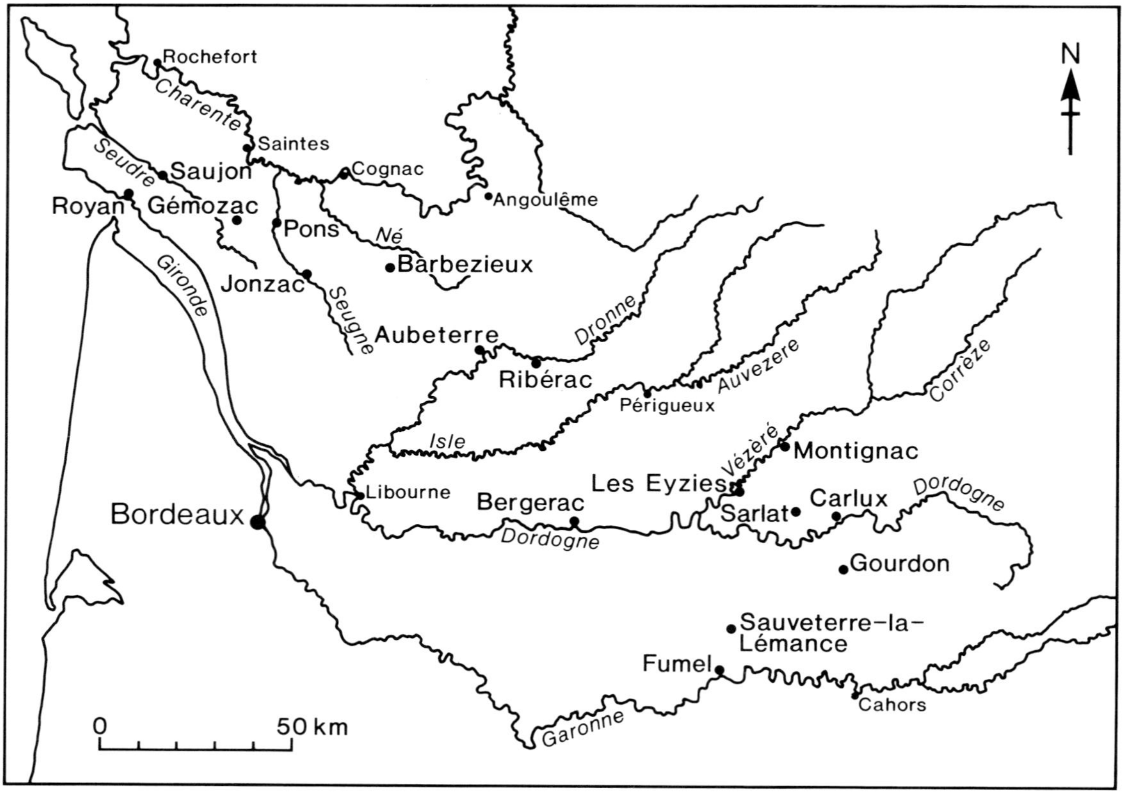

TEXT-FIG. 1. Northern Aquitaine showing some of the principal localities mentioned in the text.

Coquand cites under Troisième étage: 'trois points du département...le village de Philippeaux à l'ouest de Lamerac, les environs de Maine-Blanc, au sud de Montmoreau, et le sommet du plateau qui domine au nord de la ville d'Aubeterre' (1856, p. 94).

Van Hinte (1979) is quite wrong when he states that 'Road-side sections at Javrezac, near Cognac' form the only locality mentioned specifically by Coquand when introducing the Santonian and that 'one locality has been unambiguously singled out by Coquand: the cliff on and inside which the village of Aubeterre has been built' for the Campanian. A number of additional localities are referred to by Coquand in his later papers: the reader is referred to Séronie-Vivien (1972) for discussion.

Coquand (1858–1860) lists the faunas of the Santonian, Campanian, and Dordonian stages. The first are discussed elsewhere (Kennedy in press *a*). For the Campanian the following ammonites are noted: *Ammonites gollevillensis* d'Orbigny, *A. marroti* Coquand, *A. petrocoriensis* Coquand, *Scaphytes Nanclasi* Coquand, *S. Baylei* Coquand, *S. Heberti* Coquand, *Turrilites Archiaci* d'Orbigny, *T. Heberti* Coquand, *Baculites Faujassi* Lamarck, and *B. anceps* Lamarck.

Pachydiscus (*Pachydiscus*) *gollevillensis* (d'Orbigny, 1850) is an Upper Maastrichtian species (see revision in Kennedy 1986*a*, p. 28, pls. 1–3; pl. 4, figs. 4–6; pl. 5, figs. 12–14, 20–24; pl. 11, figs. 1–5; text-figs. 2, 3P, R, 4C); old collections contain, however, specimens of the *P.* (*P.*) *oldhami* (Sharpe, 1855)–*P.* (*P.*) *haldemsis* (Schlüter, 1867) group so labelled, and this appears to be the basis of Coquand's records. *A. marroti* is a *Hoplitoplacenticeras* (*Hoplitoplacenticeras*); the holotype survives, and is described below. *A. petrocoriensis* is a Coniacian *Forresteria* (*Harleites*) (see revision in Kennedy 1984*a*, p. 49, pl. 4, figs. 1 and 2; pl. 5, figs. 1–11; pl. 6, figs. 1–9; pl. 7, figs. 1–20; pl. 9, figs. 1–4;

pl. 21, figs. 2 and 4; text-figs. 13D–F, H, 16, 18A, B, G). *S. nanclasi* is a junior synonym of *Scaphites hippocrepis* DeKay, 1928, and is described below. The types of *S. baylei* and *S. heberti* are lost and I regard them as *nomina dubia*. *T. archiaci* is a *Nostoceras* (*Nostoceras*) described below; the type of *T. heberti* is lost, but from the description appears to be an *N.* (*Bostrychoceras*) *polyplocum* (Roemer, 1841). *Baculites faujassi* is a synonym of *B. vertebralis* Lamark, 1801, and both it and *B. anceps* are Upper Maastrichtian species (see revision in Kennedy 1986*a*). Specimens of a range of *Baculites* species in old collections are labelled *anceps* and *vertebralis*; there is no way of relating Coquand's usage to species described below.

Coquand cites no ammonites from the Dordonien, only *Nautilus Dekayi* Morton, which he records from Phillippeaux. This is a Campanian–Maastrichtian species of *Eutrephoceras* Hyatt, 1894.

The most significant nineteenth-century advances in our understanding of the Campanian and Dordonian following the work of Coquand came from the researches of Arnaud (1862–1897) and de Grossouvre (1894–1901). The former amassed enormous amounts of data and large fossil collections, much deriving from the then freshly excavated railway cuttings across the area. The latter largely based his revision of the ammonites of the 'Craie Supérieur' (de Grossouvre 1894) on Arnaud's Collection, and it formed the basis for his account of the stratigraphy of the Aquitaine Cretaceous and his zonation of Coniacian to Campanian Stages (1901).

Arnaud introduced a letter system to designate and main lithostratigraphic units in Aquitaine; their correlation with the divisions of Coquand are shown in Table 1. As I have noted elsewhere (1984*a*, p. 4) his *Mémoire sur le Terrain Crétacé du Sud-Ouest de la France* (1877*b*) is the most important synthesis, giving measured sections at many localities, precisely localizing faunas and tracing facies changes across the region. His initial division of the Craie Supérieure is as follows (1877*b*, p. 3):

CRAIE SUPÉRIEURE.

Première période: *Sénonien inférieur.*
Rivage à l'est; — Haute mer au sud.
1er série: *Coniacien.*

15. Marnes et grès: *Rhynchonella Petrocoriensis*;
16. Calcaires noduleux ou cristallins: *Ammonites tricarinatus*;
17. Calcaire glauconieux à *Rhynchonella Baugasi.*

2e série: *Santonien.*

18. Marnes et grès inférieurs: *Rhynchonella deformis*; *Botriopygus*;
19. Marnes à *Ostrea vesicularis et O. proboscidea*;
20. Marnes et gres supérieurs: *Sphaerulites Hoeninghausi*, *Ostrea acutirostris*;
21. Calcaire noduleux glauconieux; *Conoclypeus ovum.*

Deuxième période: *Campanien.*
Rivage au sud-est; — Haute mer à l'ouest.

22. Calcaires marneux hydrauliques: *Rhynchonella globata*, *Hippurites Arnaudi*;
23. Calcaire marneux arénacé: *Belemnitella quadrata*;
24. Calcaire blanc ou glauconieux: *Ananchytes ovata*, *Ostrea vesicularis major.*

Troisième période: *Dordonien.*
Rivage à l'ouest; — Haute mer au sud.

25. Calcaire glauconieux à *Orbitolites media*, *Crania Ignabergensis*, *Radiolites crateriformis*, *Sphaerulites alatus*;
26. Calcaire jaune, arénacé ou dolomitique: *Hemipneustes radiatus*, *Radiolites acuticostatus*, *Sphaerulites saemanni*;
27. Sables, poudingues dolomitiques, grès ferrugineux, avec mêmes Rudistes et Polypiers.

This is modified in the ribbon diagram at the end of Arnaud's memoir as follows:

S Grès supérieur dolomitique et poudinge marneux à Rudistes: *Hippurites Lamarcki*; *Psammechinus*

R Calcaires jaunes supérieurs: R^{b}. *Calcaires dolomitiques.* R^{a}. *Calcaires arénacés. Radiolites acuticostatus*, *Hemipneustes radiatus*, *Hemiaster prunella*

Q	Calcaire glauconieux: *Orbitoides media, Radiolites crateriformis, Rhynchopygus Marmini*	
P	Calcaires marneux à silex	*Calcaire supérieur, blanc ou glauconieux: Anachytes ovata.* *Calc. moyen, marneux ou arénacé: Belemnitella quadrata* *Calc. inférieur, marneux, hydraulique: Hippurites Arnaudi, Rhynchonella globata*
N^1	Marnes et grès Santonien supérieurs	N^o *Calcaire glauconieux à Conoclypeus ovum* Nb *Banc à Ostrea acutirostris* N^a Grès à *Sphaerulites Hoeninghausi*

Later papers include many additional details, summarized in 1878*a* where Arnaud correlated the chalks of the north of France with the Aquitaine succession as follows:

	BASSIN DU SUD-OUEST	ETAGES ET SOUS-ÉTAGE	
S	Sables, grès ferrugineux, poudingue dolomitique; *Radiolites ingens*, etc.: Beaumont-de-Périgord.	sup.	Dordonien.
R^2	Calcaires jaune-blanchâtres, tendres, à silex; *Hemiaster prunella, Cassidulus lapis-cancri*, etc.: Mussidan, Beaumont, etc.	moyen	
R^1	Calcaires dolomitiques, jaunes ou blanchâtres, avec ou sans silex, à grands Rudistes; *Hemipneustes striatoradiatus*, etc.: Meschers, Barbezieux, Aubeterre, Belvès, etc.		
Q	Calcaires marneux ou glauconieux; *Conoclypeus Leskei*: Royan, Meschers, Aubeterre: — calcaires blancs ou bleus, à silex: Neuvic; *C. Leskei*; Belvès; *C. orbicularis*; — sables à *Rhynchopygus Marmini* et *Orbitolites media*: Belvès.	inf.	
P^3	Calcaire blanc ou gris, avec cordons siliceux; *Crania Ignabergensis, Ostrea vesicularis major, O. semiplana, Ananchytes ovata, Offaster pilula, Bourgueticrinus ellipticus, etc.*: La Tremblade, Talmont, Saint-Seurin-d'Uzet, Montmoreau (Tauillard), Viville, etc.	sup.	Campanien.
P^2	Calcaire gris, marneux ou arénacé, avec cordons siliceux; *Belemnitella quadrata*; Montmoreau; — calcaire blanc ou bleu, à silex et Alvéolines: Belvès.	moyen	
P^1	Calcaire gris-bleuâtre, hydraulique, à silex: *Pyrina Petrocoriensis, Micraster glyphus, Cyphosoma Arnaudi, Terebratella Santoniensis*: Chartuzac, Livernant, Champcevinel, Trélissac, Razac, Limeyrat, La Gélie, Larzac, etc.	inf.	
N^2	Calcaire marneux, gris, à silex: Saintes; — calcaires glauconieux, gris: *Conoclypeus ovum*: Charmant, Le Bugue; — grès argileux; *Ostrea acutirostris*: Sarlat, Le Got, Villefranche.	sup.	Santonien.
N^1	Calcaire marneux; banc à *Ostrea vesicularis* et *O. proboscidea*: Granconière, Charmant, Boussitran; — calcaire à silex; mêmes Ostracées: Le Sout. Villefranche-de-Belvès.	moyen	
M^2	Calcaire noduleux: Lavalette, Périgueux; *Botriopygus Nanclasi*; — calcaire jaune arénacé, pierre de taille; *B. Toucasianus*: Miremont, Moulin-Lescot, etc.	inf.	
M^1	Calcaire marneux, gris; *Rhynchonella Eudesi*: Saint-Martin-de-Cognac, Périgueux, Miremont; — calcaire rouge compacte: La Trape, Saint-Cernin, Freycinet-le-Gelat, etc.		

It is important to note that Arnaud modified the Dordonien as proposed by Coquand. As originally conceived this included only the beds with *Hippurites radiosus*, *Sphaerulites cylindraceus*, and *Radiotites Jouanneti* (Coquand's nomenclature), placed in Assise R^1 by Arnaud. He proposed that the faunal affinities of Assise Q below argued for the placement of this unit in the Dordonien. It is also necessary to note that Assise S is not Cretaceous (J. P. Platel, *pers. comm.*).

De Grossouvre (1894) retained the letter system of Arnaud, but reverted to the Campanien–Dordonien division of Coquand, treating Assise Q of Arnaud as Campanien supérieur. De Grossouvre (1901) described the Santonian to Dordonian of Aquitaine in some detail (pp. 375–383), listing the ammonite faunas in sequence. In the discussion of the Upper Cretaceous in general (Chapter 22) de Grossouvre recognized four ammonite zones in the Campanian, as discussed below;

the terms Dordonien and Maestrichtien were abandoned, with the Campanien extending to the top of the Cretaceous.

The next significant work on the Aquitaine sequence is Séronie-Vivien's revision of the Coniacian, Santonian, and Campanian stages in Aquitaine (1956, 1960, 1972). Her major synthesis (1972) provides a full bibliography and discussion of intervening studies, lists some additional ammonite occurrences, but in general applies the nomenclature of de Grossouvre and Arnaud. This work stresses the microfacies and microfaunas of the sequence and of interest to the present discussion in the usage of the terms Dordonian and Maastrichtian in Aquitaine. The former is regarded as equivalent to a part of the latter, and the base of the Maastrichtian in Aquitaine is taken at the appearance of the foraminifer *Orbitoides media*. The base of the Dordonian *sensu* Arnaud and Maastrichtian *sensu* Séronie-Vivien thus correspond.

Subsequent work by M. Neumann, J. P. Platel, and their colleagues has produced a series of papers on the microfaunas and microfloras of the Upper Cretaceous in Aquitaine. Most significant for the present study is the subdivision of the succession in Grand Champagne by Platel (1977) who has mapped much of the area. He reviewed previous work and recognized five lithological divisions in the Campanian, plus eight microfaunal assemblage zones, CI–CVIII. Text-fig. 2 shows his divisions and their correspondence with the generalized lithological succession in the area, while Table 1 compares earlier litho- and biostratigraphic divisions.

It will be noted that the Dordonian is regarded as Upper Campanian, and that no Maastrichtian is recognized. The reader is referred to this excellent study for a discussion of the problems associated with the various microfossil schemes used in Aquitaine.

More recently Neumann (1980*a*, *b*, *c*), Lambert (1980, 1981), Neumann and Platel (1985*a*), and Neumann *et al.* (1984) have presented syntheses integrating microfaunal and microfloral distributions, while the last mentioned publication suggests that there is indeed Maastrichtian present in the northern part of the Aquitaine Basin, recording detailed borehole successions near Aubeterre (Charente) and Maurens (Dordogne). Neumann and Platel (1985*b*) provide a synthesis of ammonite occurrences in northern Aquitaine based on citations in de Grossouvre and Arnaud's publications, linking them to Platel's (1977) zonal scheme. Since they did not study the specimens or sample matrix from existing material this survey is of limited value.

Much additional data on other fossil groups is to be found in the proceedings of the *Colloque sur les étages Coniacien à Maastrichtien, Marseille 26–28 Septembre, 1983* (*Géologie Méditerraneenne*, **10**, 3–4, 1985).

LIMITS OF THE CAMPANIAN STAGE

In Aquitaine the boundary between Campanian and Santonian Stages is gradational so far as the lithological succession is concerned. The typical facies of the top Santonian in northern Aquitaine is a grey-green bryozoan-rich chalk with white silicifications ('morilles' of French workers), the lowest Campanian is grey marly chalk with *Pycnodonte vesicularis*. Microfaunas in the region provide the best local basis for definition, notably the benthic foraminifer assemblages: see Platel (1977) and Neumann and Platel (1985*a*) for details. Bellier (in Neumann and Platel Coordonnateurs 1985*a*) notes the appearance of the planktonic foraminifer *Globotruncana* cf. *arca* (Cushman) as an indicator of proximity to the base of the stage. Neumann and Platel (1985*a*) place *Pachydiscus* (*Pachydiscus*) *launayi* de Grossouvre, 1894, *P.* (*P.*) *duelmensis* (Schlüter, 1872*a*), *Eupachydiscus isculensis* (Redtenbacher, 1873), and *Placenticeras bidorsatum* (Roemer, 1841) in benthonic foraminifer Zone CI, regarding this as the lowest ammonite fauna in the type region, a conclusion supported here.

Ammonite workers have usually taken a *P. bidorsatum* Zone as marking the base of the Campanian. This can certainly be applied to the type area and the species also occurs in Westphalia and Lower Saxony in the German Federal Republic. This limited distribution (plus the fact that there are to my knowledge no sections where the species can still be collected) indicate that however strong

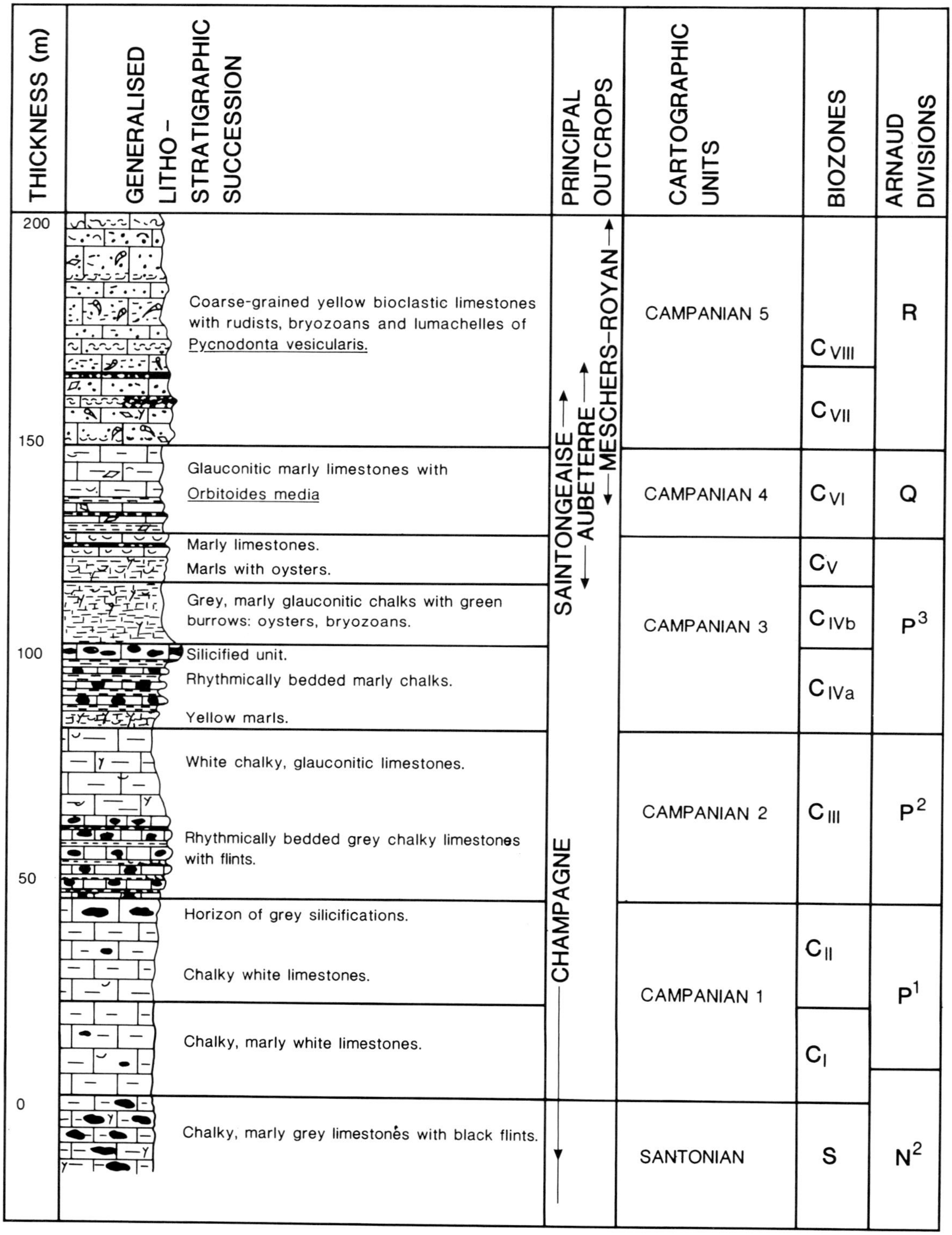

TEXT-FIG. 2. The Campanian succession in Grand Champagne showing the lithological succession, extent of principal sections, cartographic and microfaunal divisions compared with the letter system of Arnaud. Modified after Platel (1977).

the historical case may be for so defining the stage, it has little practical value and that a working standard must be based on a first-order correlation of the first appearance of *P. bidorsatum* and some other taxon. The best dating of this species in the White Chalk sequences of north Germany is given by Ernst and Schmid (1979, fig. 11) who show it first occurring just above the base of the *Gonioteuthis granulataquadrata* Zone in the Braunschweig–Hannover area. This is of great significance, because the base of this Zone is widely taken as the base of the Campanian in the Boreal Realm, and corresponds to the extinction point of *Marsupites*, widely taken to mark the top of the Santonian (see discussion in Birkelund *et al.* 1984; Schulz *et al.* 1984).

There are two other potential ammonite groups that might define the boundary, the *Scaphites hippocrepis* (DeKay, 1828) lineage and the *Texanites* → *Submortoniceras* lineage, specifically the appearance of *S. spathi* Collignon, 1948 (Kennedy 1984*b*; Birkelund *et al.* 1984). The *hippocrepis* lineage is described in great detail by Cobban (1969). The attraction of this group is its occurrence in the US western interior, Gulf Coast and Eastern Seaboard regions, the White Chalk and other facies of western Europe and northern Aquitaine, while closely allied species occur in Madagascar and Zululand: it thus spans several late Cretaceous biotic regions.

The lineage first appears in the US western interior, where three successive forms of *Scaphites leei* Reeside, 1927 are succeeded by three forms of *S. hippocrepis* DeKay, 1828. Of these, only *S. hippocrepis* III and passage forms to *S. hippocrepis* II occur outside the US western interior. Cobban places his locally defined Santonian boundary between *S. leei* III and *S. hippocrepis* I. In Aquitaine, forms corresponding to *S. hippocrepis* III occur in Assise P^2. A large assemblage from Assise P^1 *sensu lato* includes forms with characters of both *S. hippocrepis* II and III. It is important to note, however, that de Grossouvre (1901, p. 378) states that *S. hippocrepis* and *aquisgranensis* (the latter a synonym) first appear in his Assise P^{1b} a little above the first appearance of *Placenticeras bidorsatum* in his Assise P^{1a} in northern Aquitaine, and are thus lower, but not lowest Campanian. In the Vaals Formation of the Dutch–Belgian border area *S. hippocrepis* III occurs with *Pachydiscus* (*Pachydiscus*) *duelmensis*, a species known only from the lowest Campanian in northern Aquitaine. The most precise records in Germany are those of Schmid and Ernst (1975) who record *S. hippocrepis* III from the upper part of their Lower Campanian, above the disappearance of *P. bidorsatum* (Ernst and Schmid 1979, fig. 11). In Europe at least, the appearance of *S. hippocrepis* thus seems to be some way above the base of the Campanian, possibly in the lower part of the *G. quadrata* Zone in Germany, and certainly in the upper part of that zone in both England and Germany (Schmid and Ernst 1975; Gale 1980).

There is only one unlocalized *Submortoniceras* known from Aquitaine; none are known associated with *P. bidorsatum* anywhere. In Texas *S. hippocrepis* III co-occurs with *Submortoniceras*, although first appearing some way above the first appearance of the latter (Young 1963, fig. 4). In ammonite terms, *P. bidorsatum* thus appears before *S. hippocrepis* II–III in Europe, and use of a *hippocrepis* Zone at the base of the stage (e.g. Wiedmann 1979) is misleading.

The correlation of the base of the Maastrichtian between the boreal belemnite successions of the White Chalk facies of the Anglo-Paris Basin and the succession in Aquitaine is a source of great contention. The ammonite fauna of the Maastrichtian in its type area is revised elsewhere (Kennedy 1986*b*, in press *b*) and as is now well known, it is mostly Upper Maastrichtian. There is no agreed ammonite zonation for the Maastrichtian in northern Europe: it is belemnites that provide the best zonal system.

The standard scheme is as follows:

Upper Maastrichtian	*Belemnella casimirovensis* *Belemnitella junior*
Lower Maastrichtian	*Belemnella occidentalis* *Belemnella lanceolata*
Upper Campanian	*Belemnitella langei* *Belemnitella minor* *Belemnitella mucronata*

Schulz (1979) working in the White Chalk of north Germany further divided the Lower Maastrichtian into six *Belemnella* Zones:

Belemnella occidentalis	*Belemnella fastigiata* *Belemnella cimbrica* *Belemnella sumensis*
Belemnella lanceolata	*Belemnella obtusa* *Belemnella pseudobtusa* *Belemnella lanceolata*

The Maastrichtian stratotype in the ENCI quarry below Lichtenberg Farm (Dumont 1849; Mourlon 1878; see Romein 1962 for a more readily accessible account) is within the Upper Maastrichtian *junior* Zone. Elsewhere in the area the work of Schulz and Schmid (1983) shows that the lower part of the Maastrichtian is very incomplete. The generally agreed base to the stage is at the base of the *B. lanceolata* Zone in the White Chalk facies; this is very close to the classic boundary in ammonite terms, which is marked by the appearance of *Hoploscaphites constrictus* (J. Sowerby 1817). Thus at Kronsmoor in northern Germany this species first appears 3·5–5·0 m above the base of the *lanceolata* Zone (Surlyk 1975, 1982; Schulz 1978).

This boundary is hard to correlate with the Aquitaine succession. *Bostrychoceras polyplocum*, a *Belemnitella minor* Zone species in north Germany (Schmid and Ernst 1975) occurs in Assizes P^3 and Q in Aquitaine, that is to say benthonic foraminifer Biozones CV (*fide* Neumann and Platel 1985*b*, p. 119) and CVI of Platel (1977). Undoubted Campanian *Pachydiscus* (*Pachydiscus*) of the *oldhami–haldemsis* group occur in P^3 and Q according to Arnaud's surviving specimens. Much has been made of records of *B. mucronata* (Link, 1807) in northern Aquitaine. Arnaud (1896) recorded a specimen from the equivalent of Assise Q at Barret, near Barbezieux (Charente) and Gillard (1943) notes specimens from the Font-de-Cherves borehole no. 2 at Royan (Charente-Maritime), 1·5 km east of St-Léger-en-Pons (Charente-Maritime), south-west of Lonzac (Charente-Maritime), 2·4 km south-east of Juillac-le-Coq (Charente-Maritime), 1·5 km south-east of Plassac (Gironde), Lamérac and Maine-Blanc (Charente), and Basac (Charente). I have not seen these, or indeed any of the belemnites recorded from Aquitaine, and I regard the validity of the determinations as open to question. The species is typically lower Upper Campanian, and identification from Assise Q must be questioned, as must those from Saint Leger and Plassac, which are in the Lower Campanian according to Neumann and Platel (1985*b*).

The highest Cretaceous preserved in the northern part of the Aquitaine Basin outcrops around Maurens-le-Bost (Dordogne); it was described by Arnaud (1892) and has been reinvestigated by Neumann *et al.* (1984). Their summary of the succession is shown in text-fig. 3, which is based upon a new borehole at Le Bost. Madame Neumann has kindly examined the matrix of ammonites from this succession, including those mentioned by Arnaud (1892), some of which were described by de Grossouvre (1894), as well as those collected subsequently.

Bed 6 of the Maurens sections is correlated with the level of *Hippurites radiosus* and other rudists, and is thus equivalent to the Dordonian in the sense of Coquand's original definition. The paralectotype of *Sphenodiscus ubaghsi* de Grossouvre, 1894 (p. 142, pl. 9, fig. 6) is from Bed *b* of Arnaud (1892), the matrix showing it to be from the equivalent of Beds 7 or 8 of Neumann *et al.* (1984) and thus below the Dordonian of Coquand. The lectotype was collected, probably by de Grossouvre, from the 'Calcaire gréseaux à Thécidees' at Maurens (1894, p. 142, pl. 9, fig. 4), which I take to be Bed *h* of Arnaud (1892), the only unit where he records thecidacean brachiopods. This in turn corresponds to Bed 3 of Neumann *et al.* (1984), above the Dordonian of Coquand. Specimens of *Baculites leopoliensis* Nowak, 1908 from Maurens and in the Arnaud Collection are labelled *B. anceps*? Lam. R^2. Arnaud (1892) records *anceps* only from his Bed *h* around Maurens, corresponding to Bed 3 of Neumann *et al.* (1984). The matrix of one specimen confirms derivation from Bed 3.

Arnaud's *Bostrychoceras polyplocum*, recorded from his Bed *h*, correlated with Bed 3 of Neumann *et al.* (1984) is here identified as *Nostoceras* (*Nostoceras*) *hyatti*. Neumann *et al.* (1984) thought it more likely that this specimen came from their Bed 5, but examination of the matrix revealed debris of

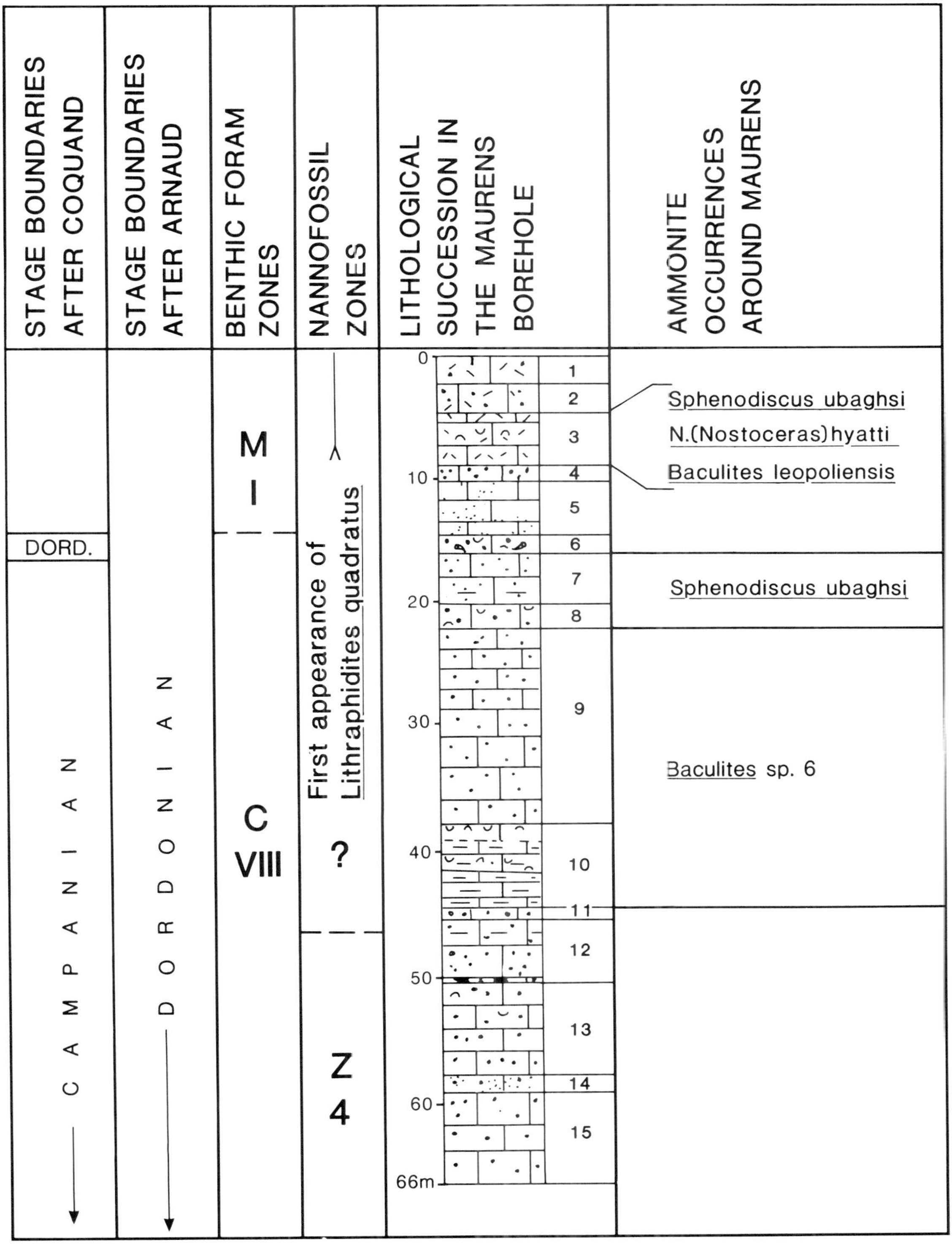

TEXT-FIG. 3. The Campanian–Maastrichtian sequence at Maurens-le-Bost (Dordogne). Modified after Neumann *et al* 1984.

diverse larger benthic foraminifers including *Lepidorbitoides*, all suggesting derivation from Bed 3 of Neumann *et al.* This evidence is tantalizingly inadequate for precise correlation with the White Chalk sequence. Only *Baculites leopoliensis* is known from northerly sections, being first described from the Lower Maastrichtian of Lipnicki, formerly in Galicia, but now in the Ukraine: I have no data on its precise age in this region. Recent work in the Petites-Pyrénées (Kennedy *et al.* 1986) revealed this species in the niveau de transition between the Marnes de Plagne and the Calcaire nankin in the Plagne and Saint Martory–Saint Marcet anticlines, the same unit also yielding *S. ubaghsi* and *Pseudokossmaticeras tercense* (Seunes, 1890). The underlying Marnes de Plagne are 3000 m thick and span the Campanian–Maastrichtian boundary in foraminifer terms, with *Globotruncana stuartiformis* Zone faunas succeeded by those of the *G. contusa* Zone. 50 m from the summit *Bolivinoides draco-miliaris* occurs, and the upper part yields *Globotruncana contusa*, *G. gansseri*, *G. trinidadensis*, and *G. ventricosa* (Segura 1979; Bilotte 1985). There is thus no doubt of the age of the succeeding niveau de transition in foraminifer terms. This is supported by the presence of *P. tercense*, regarded as lower Lower Maastrichtian in Bulgaria, the USSR, and Carinthia, Austria. In the Maurens section there is thus no doubt that Bed 3 of the section given in text-fig. 3 is already Lower Maastrichtian in foraminifer terms. *S. ubaghsi* is Lower Maastrichtian in the same terms in the Petites-Pyrénées, but there is no evidence to suggest that its first appearance is within the Maastrichtian. Thus its occurrence in Bed 7 or 8 at Maurens could be Upper Campanian or Lower Maastrichtian. Bed 6 at Maurens, corresponding to the rudist horizon that is Coquand's Dordonian could thus be Campanian or Maastrichtian on the basis of the currently available ammonite evidence. I remain unclear as to how this microfossil-based Campanian–Maastrichtian boundary relates to the belemnite standard.

STRATIGRAPHIC DISTRIBUTION OF THE FAUNA

As noted in the introduction, ammonites are so rare in the Campanian at the present day that I have based this account almost entirely upon nineteenth-century collections. As a result, vertical distribution is discussed in terms of Arnaud's letter divisions, based upon his surviving collections and detailed records and additional records and observations in de Grossouvre (1894, 1901). As I have noted elsewhere (Kennedy 1984*b*, 1985) the faunal assemblages listed by de Grossouvre do not form a zonal succession in the commonly understood sense. Ammonites (other than *Baculites*) are restricted to narrow horizons, separated by barren chalk. The standard succession is:

Upper Campanian	*Bostrychoceras polyplocum* Zone
	Hoplitoplacenticeras vari Zone
Lower Campanian	*Menabites (Delawarella) delawarensis* Zone
	Diplacomoceras bidorsatum Zone

Taxonomic revision below modifies the nomenclature of these zonal indices, but the index species are all rare. I have seen 13 specimens of *Placenticeras bidorsatum*, 2 of *M.* (*D.*) *delawarensis* (and only 4 *Menabites* (*Delawarella*) in total), 25 *H. marroti* (the correct name for Aquitaine *vari*), and 14 *Nostoceras* (*Bostrychoceras*) *polyplocum*. With this scarcity in mind, and an equal rarity of other species it is not possible to refine the scheme further, or introduce alternative indices. The succession summarized in Table 2 and discussed below is thus quite different from that developed for the Cenomanian to Coniacian Stages. The most one can ever expect in northern Aquitaine is to date the ammonite occurrences in terms of the detailed benthic foraminifer zones of Platel (1977) in order ultimately to correlate the sequence with the belemnite–echinoderm–inoceramid dominated successions of northern Europe, and the planktonic foraminifer and coccolith sequences of the Tethys.

The lowest ammonite assemblage recognized comes from the lower part of Assise P^1, P^{1a} of de Grossouvre (1901), and is the *P. bidorsatum* Zone (the lower part). It includes *P. bidorsatum*, *Pachydiscus* (*Pachydiscus*) *duelmensis* (Schlüter, 1872*a*), *P.* (*P.*) *launayi* de Grossouvre, 1894, and *Eupachydiscus isculensis* (Redtenbacher, 1873). Above, the upper part of the *Placenticeras bidorsatum* Zone, P^{1b} of de Grossouvre (1901), yields *P. bidorsatum* and *Scaphites* (*Scaphites*) *hippocrepis* (DeKay, 1828). De Grossouvre (1901, p. 378) records *Pachydiscus levyi* de Grossouvre, 1894, but I cannot

TABLE 2. Stratigraphic distribution of Campanian and Lower Maastrichtian ammonites in northern Aquitaine.

	P^1	P^2	P^3	Q	R^1	R^2
Pseudophyllites indra (Forbes, 1846)			×			
Yokoyamoceras aff. *jimboi* Matsumoto, 1955	×					
Hauericeras fayoli de Grossouvre, 1894				?	?	?
Hauericeras sp.			×			
Brahmaites brahma (Forbes, 1846)						×
Pachydiscus (Pachydiscus) duelmensis (Schlüter, 1872*a*)	×					
P. (Pachydiscus) haldemsis (Schlüter, 1867)			×			
P. (Pachydiscus) colligatus (Binkhorst, 1861)				×		
P. (Pachydiscus) oldhami (Sharpe, 1855)				×		
P. (Pachydiscus) launayi de Grossouvre, 1894	×	×				
P. (Pachydiscus) cf. *complanatus* Collignon, 1952		×				
P. (Pachydiscus) sp.			×			
Eupachydiscus isculensis (Redtenbacher, 1873)	×					
Praemuniericeras sp.	?	?	?			
Placenticeras bidorsatum (Roemer, 1841)	×	×	×			
Hoplitoplacenticeras (Hoplitoplacenticeras) marroti (Coquand, 1859)		×	×			
H. (Hoplitoplacenticeras) coesfeldiense (Schlüter, 1867)			?			
H. (Hoplitoplacenticeras) dolbergense (Schlüter, 1876)			×			
H. (Hoplitoplacenticeras) rejaudryi de Grossouvre, 1894			×			
Submortoniceras sp.	?	?	?			
Menabites (Delawarella) delawarensis (Morton, 1830)	×					
M. (Delawarella) campaniense (de Grossouvre, 1894)	×	?				
M. (Australiella) cf. *besairiei* (Collignon, 1938)	?	?				
Sphenodiscus ubaghsi de Grossouvre, 1894					×	×
Sphenodiscus sp.						×
Nostoceras (Nostoceras) archiacianum (d'Orbigny, 1842)				×		
N. (Nostoceras) hyatti (Stephenson, 1941)						×
N. (Bostrychoceras) polyplocum (Roemer, 1841)			×	×		
Neocrioceras (Schlueterella) ? *elegans* (Moberg, 1885)			×			
Neancyloceras cf. *bipunctatum* (Schlüter, 1872*a*)			×			
Neoglyptoxoceras (?) *retrorsum* (Schlüter, 1872*a*)	?	?	×			
Pseudoxybeloceras (Parasolenoceras) cf. *interruptum* (Schlüter, 1872*a*)				×		
Baculites sp. 1		×	×			
Baculites sp. 2	×	×				
Baculites sp. 3				×		
Baculites sp. 4		×				
Baculites sp. 5			×			
Baculites sp. 6						×
B. leopoliensis Nowak, 1908						×
Baculites sp. 7						×
Scaphites (Scaphites) haugi de Grossouvre, 1894			×			
S. (Scaphites) hippocrepis (DeKay, 1828)	×	×				
S. (Scaphites) sp.		×				
S. (Scaphites) gibbus Schlüter, 1872*a*			×			
Trachyscaphites spiniger Schlüter, 1872*a*			×			
T. pulcherrimus (Roemer, 1841)			×	×		

confirm this record. Specimens of *Menabites* from P never occur with *S. (S.) hippocrepis*, and I believe them to be from a higher horizon, as did de Grossouvre.

The succeeding *M. (D.) delawarensis* Zone yields few ammonites. I believe all the *Menabites* from Aquitaine to be from this level: *M. (D.) delawarensis* (Morton, 1830), *M. (D.) campaniense* (de

Grossouvre, 1894), *M.* (*Australiella*) cf. *besairiei* (Collignon, 1938). It is not possible to assign other species definitely to this zone on available information.

The *H. marroti* Zone yields relatively numerous specimens of the index species. There is a specimen of *Placenticeras bidorsatum* in the Arnaud Collection labelled 'P^3, Tauillard', which is much younger than other specimens of the species, if it is correctly horizoned. *Baculites* sp. 2 is undoubtedly from this level, as is *S.* (*S.*) *gibbus* Schlüter, 1872*a* and *S.* (*S.*) *haugi* de Grossouvre, 1894. Some other species recorded from P^3 may be from this or the succeeding *N.* (*Bostrychoceras*) *polyplocum* Zone. The latter yields, in addition to the index species, *Pachydiscus* (*Pachydiscus*) *colligatus* (Binkhorst, 1861), *P.* (*P.*) *oldhami* (Sharpe, 1855), *N.* (*Nostoceras*) *archiacianum* (d'Orbigny, 1842), *Neocrioceras* (*Schlueterella*)? *elegans* (Moberg, 1885), *Neancyloceras* cf. *bipunctatum* (Schlüter, 1872*a*), *Neoglyptoxoceras retrorsum* (Schlüter, 1872*a*), *Pseudoxybeloceras* (*Parasolenoceras*) cf. *interruptum* (Schlüter, 1872*a*), *Baculites* sp. 2 and 3, *Trachyscaphites spiniger* Schlüter, 1872*a*, and *T. pulcherrimus* (Roemer, 1841).

Above, the next ammonite record is from Beds 7 or 8 of the Maurens section, a single *Sphenodiscus ubaghsi* de Grossouvre, 1894, which may be high Upper Campanian or Lower Maastrichtian, as already discussed. The highest *in situ* fauna is that of Bed 3 of the Maurens section, corresponding to Assise R^2 of Arnaud. The fauna is regarded as Lower Maastrichtian on the basis of microfossil data, with *S. ubaghsi* de Grossouvre, 1894, *Nostoceras* (*Nostoceras*) *hyatti* Stephenson, 1941, and *Baculites leopoliensis* Nowak, 1908. No formal zonal name is applied to this assemblage at this time, nor is its position in relation to the standard Boreal belemnite sequence unequivocally established.

CONVENTIONS

Location of specimens. The following abbreviations are used to indicate the location of specimens mentioned in the text: BMNH, British Museum (Natural History), London; EMP, École des Mines Collections, formerly in Paris but now in the Université Claude-Bernard, Lyon; FSL, Faculté des Sciences, Université Claude-Bernard, Lyon; GBA, Geologische Bundesanstalt, Vienna; GK, Department of Geology, Kyushu University; GPIB, Geol. Pal. Institüt, Bonn; GSM, Geological Museum, including the Collections of the Institute of Geological Sciences, London; GT, Department of Geology, Tokyo University; IRSNB, Institut Royal des Sciences Naturelles, Brussels; MAFI, Hungarian Geological Survey Collections, Budapest; MNB, Museum für Naturkunde, Berlin; MMK, Geologisk Museum, Copenhagen; MNHP, Muséum National d'Histoire Naturelle, Paris; NHMW, Naturhistorisches Museum, Vienna; OUM, University Museum, Oxford; SP, Collections of the Sorbonne, now in the Université Pierre et Marie Curie, Paris.

Suture terminology. The system of Wedekind (1916) as reviewed by Kullmann and Wiedmann (1970) is used here. E = external lobe, L = lateral lobe, U = umbilical lobe, I = internal lobe.

Dimensions. All dimensions are given in millimeters; D = diameter, Wb = whorl breadth, Wh = whorl height, and U = umbilicus; c = costal and ic = intercostal. Figures in parentheses refer to dimensions as a percentage of diameter. The term *rib index* as applied to heteromorphs is the number of ribs in a distance equal to the whorl height at the mid-point of the interval counted.

Synonymies. Only citations which include illustrations of material or important systematic, stratigraphic, or geographic information are included.

SYSTEMATIC PALAEONTOLOGY

Order AMMONOIDEA Zittel, 1884, pp. 355, 392
Suborder LYTOCERATINA Hyatt, 1889, p. 7
Superfamily TETRAGONITACEAE Hyatt, 1900, p. 568
Family TETRAGONITIDAE Hyatt, 1900, p. 568
Genus PSEUDOPHYLLITES Kossmat, 1895, p. 137 (41)

Type species. Ammonites indra Forbes, 1846, p. 105, pl. 11, fig. 7, by original designation.

Discussion. See Kennedy and Klinger (1977, p. 180) and Henderson and McNamara (1985, p. 49) for diagnosis and discussion. There are three main species in this genus: *Pseudophyllites indra* (Forbes,

1846), *P. teres* (Van Hoepen, 1920), and *P. latus* (Marshall, 1926), differentiated on details of coiling, whorl section, and suture. *Pseudophyllites* are adult at diameters of up to 300 mm, far in excess of earlier and contemporary tetragonitid genera; it appears to be a hypermorphic giant.

Occurrence. Typically Lower Campanian to Upper Maastrichtian but recorded from the Santonian of Madagascar (Collignon 1956). It has a near world-wide distribution: Antarctica, South Africa, Madagascar, south India, Australia, New Zealand, Japan, Sakhalin, southern and central Europe, northern Spain, Landes and Aquitaine in France, Poland, Austria, Northern Ireland, west Greenland, Alaska, British Columbia, California, US Gulf Coast and Atlantic Seaboard, Brazil, and Chile.

Pseudophyllites indra (Forbes, 1846)

Plate 1, figs. 1–5; text-figs. 4E, 5A, 6A–E

1846 *Ammonites indra* Forbes, p. 105, pl. 11, fig. 7.
1977 *Pseudophyllites indra* Forbes, 1846; Kennedy and Klinger, p. 182, figs. 19A–F, 20–22 (with full synonymy).
1980 *Pseudophyllites indra* (Forbes); Błaszkiewicz, p. 19, pl. 56, figs. 4 and 5.
1985 *Pseudophyllites indra* (Forbes, 1846); Henderson and McNamara, p. 50, pl. 2, figs. 7 and 8; pl. 3, figs. 4 and 5; text-fig. 5*a*, *d*.

Lectotype. By subsequent designation of Kennedy and Klinger 1977, p. 182, BMNH C51068, the original of Forbes 1846, pl. 11, fig. 7, from the Valudayur Group near Pondicherry, south India (text-fig. 6D–E).

Material. SP unregistered *ex* Arnaud Collection from Assise P^3, Petingaud (Charente).

Dimensions.

D	*Wb*	*Wh*	*Wb*:*Wh*	*U*
63·5(100)	31·0(48·8)	32·5(51·2)	0·95	12·6(19·8)
99·0(100)	49·0(49·5)	51·2(51·7)	0·96	18·3(18·5)

Description. The specimen is a wholly septate composite internal mould. Coiling involute, 66% of previous whorl covered. Umbilicus small, deep, conical with flattened, outwards-inclined umbilical wall. Whorl section slightly compressed, greatest breadth around mid-flank at 60 mm diameter, close to umbilical shoulder at 100 mm. Umbilical wall separated from broadly rounded, convergent flanks by sharply defined umbilical shoulder; venter very broadly rounded. No ornament preserved. External suture ill-preserved, U slightly retracted, L bifid. Internal suture with massive septal lobe.

Discussion. The specimen closely resembles the lectotype but for the absence of ornament: this reflects the poor preservation. *Pseudophyllites latus* (Marshall, 1926) (p. 149, pl. 20, fig. 6; pl. 32, figs. 1 and 2) (see also Kennedy and Klinger 1977, p. 190, figs. 25 and 26; Henderson and McNamara 1985, p. 50, pl. 3, figs. 1–3, 6–8; text-fig. 5*b*, *c*) of which *whangaroaensis* Marshall, 1926 (p. 153, pl. 20, fig. 2; pl. 21, fig. 11, pl. 32, figs. 5 and 6), *peregrinus* Spath, 1953 (p. 7, pl. 1, figs. 6–9), and *skoui* Birkelund, 1965 (p. 37, pl. 3, figs. 2–6; text-figs. 26–33) are junior synonyms and *Desmoceras loryi* Kilian and Reboul, 1909 (p. 18, pl. 1, figs. 4 and 5) a possible senior synonym, has broader whorls, less rapidly increasing whorl height, and a much steeper umbilical wall (*fide* Henderson and McNamara 1985, text-fig. 5). There are also sutural differences, the ventral saddle being lanceolate in *indra* and spatulate in *latus* (Henderson and McNamara 1985, text-fig. 5*c*, *d*) while the septal lobe and recess is large in *indra* and small in *latus*. *P. teres* (Van Hoepen, 1920) (p. 144, pl. 25, figs. 1 and 2) has compressed whorls with subparallel sides and a subvertical umbilical wall. It seems to appear in the Santonian and to be a small species, in some respects transitional to *Tetragonites* Kossmat, 1895. *Takahashia* Matsumoto, 1984*a* (p. 33, fig. 1) with *T. eureka* Matsumoto, 1984*a* as type species was described as a heterochronous homoeomorph of *Pseudophyllites* with shell form and suture of *P. indra* but differing in the presence of radial ribs in addition to fine radial lirae and delicate spiral lirae. As will be seen from text-fig. 6D, E the lectotype of *indra* possesses all these features and the two differ in strength of ribbing only.

Gaudryceras colloti de Grossouvre, 1894 (p. 229, pl. 37, fig. 8) from the Upper Maastrichtian Calcaire à *Stegaster* between Gan and Rébénacq (Pyrénées-Atlantiques, France) was regarded as a

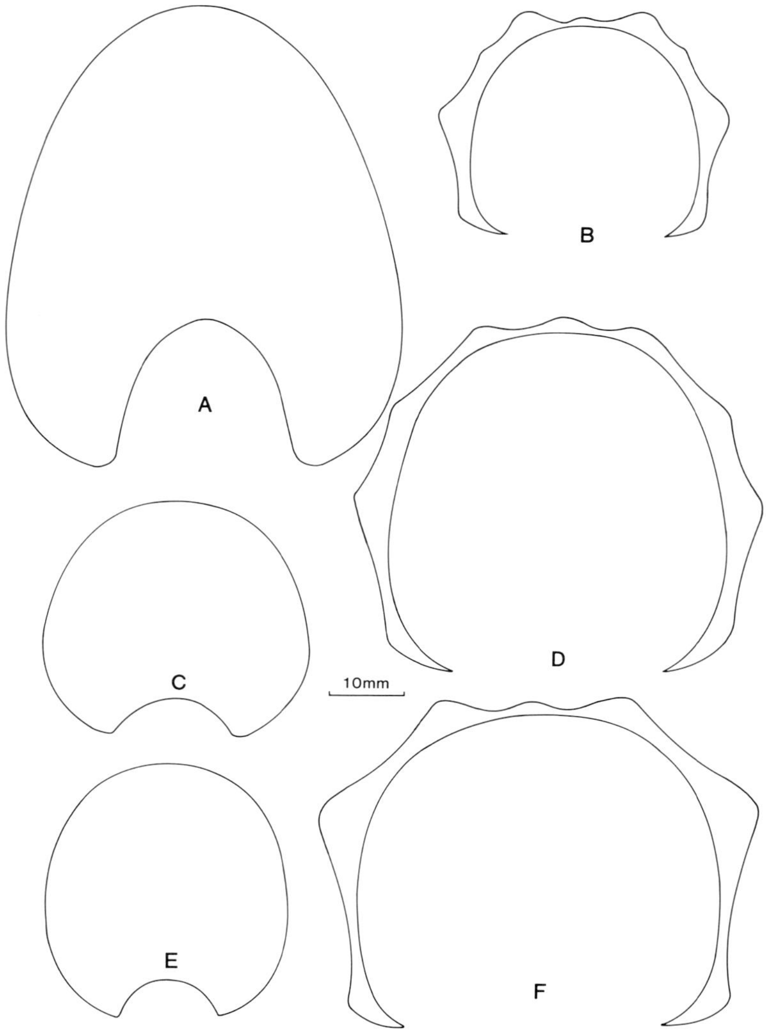

TEXT-FIG. 4. Whorl sections of: A, *Pachydiscus oldhami* (Sharpe, 1855); B, D, *Menabites* (*Delawarella*) *campaniense* (de Grossouvre, 1894); C, *P.* (*Pachydiscus*) *launayi* de Grossouvre, 1894; E, *Pseudophyllites indra* (Forbes, 1846); F, *Menabites* (*Australiella*) cf. *besairiei* (Collignon, 1938).

EXPLANATION OF PLATE 1

Figs. 1–5. *Pseudophyllites indra* (Forbes, 1846). SP unregistered *ex* Arnaud Collection, from Assise P^3, Petingaud (Charente).

Figs. 6–8. *Yokoyamoceras* aff. *jimboi* (Matsumoto, 1955). SP unregistered *ex* Arnaud Collection, from Assise P^1, Verteillac (Dordogne).

Fig. 9. *Menabites* (*Delawarella*) *campaniense* de Grossouvre, 1894. SP unregistered *ex* Arnaud Collection, a paralectotype, from Assise P^2, Montmoreau (Charente).

All figs. $\times 1$.

KENNEDY, *Pseudophyllites*, *Yokoyamoceras*, *Menabites* (*Delawarella*)

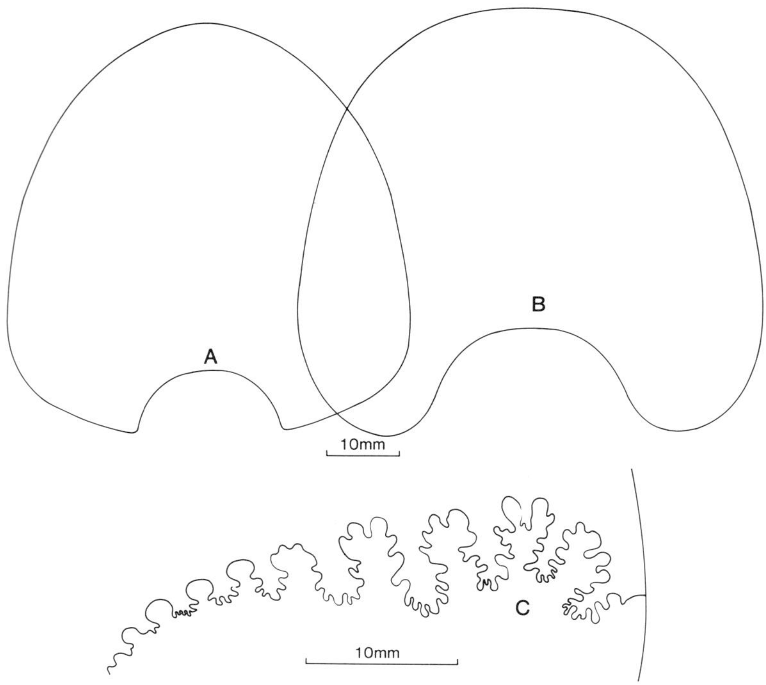

TEXT-FIG. 5. Whorl sections of: A, *Pseudophyllites indra* (Forbes, 1846); B, *Pachydiscus* (*Pachydiscus*) *launayi* de Grossouvre, 1894. External suture of: C, *Sphenodiscus ubaghsi* de Grossouvre, 1894.

possible *P. indra* by Kennedy and Klinger (1977). I have since seen the figured specimen (EMP unregistered): it is a well-preserved (if crushed) composite internal mould of a juvenile *indra* (text-fig. 6A–C).

Ammonites anaspastus Redtenbacher, 1873 (p. 113, pl. 26, fig. 1), referred to *Mesogaudryceras* Spath, 1927 by Reyment (1958, p. 32, who refigured the holotype: pl. 9, fig. 1) from the Maastrichtian of Neuberg, Steiermark, Austria, proves to be an ill-preserved *P.* cf. *indra* (Kennedy and Summesberger 1986).

Occurrence. (Upper Santonian?) Lower Campanian to Upper Maastrichtian. Assise P^3 of the Campanian in Aquitaine, Upper Maastrichtian of the Pyrénées-Atlantiques, Campanian or Maastrichtian of Tercis (Landes) (EMP Collections), Upper Campanian of Northern Ireland and Poland, Maastrichtian of Neuberg, Steiermark, Austria. Also known from south India, Zululand (South Africa), Madagascar, Western Australia, Japan, Sakhalin, Alaska, British Columbia, California, and, perhaps, Brazil. Dr. W. A. Cobban (pers. comm.) has specimens from New Jersey and Mississippi, USA.

Suborder AMMONITINA Hyatt, 1889, p. 7
Superfamily DESMOCERATACEAE Zittel, 1895, p. 426

[*nom. transl.* Wright and Wright, 1951, p. 18; *ex* Desmoceratidae Zittel, 1895]

Family DESMOCERATIDAE Zittel, 1895, p. 426
Subfamily PUZOSIINAE Spath, 1922*a*, p. 126

[= Silesitoidinae Breistroffer, 1953, p. 74; Melchioritinae Beskovski, 1977, p. 892 (*nom. correct.* Wright and Kennedy 1984, p. 54 *ex* Melcioritinae Beskovski); Abrytusitinae Beskovski, 1977, p. 893 (*nom. correct.* Wright and Kennedy 1984, p. 54 *ex* Abritusitinae Beskovski)]

Genus YOKOYAMOCERAS Wright and Matsumoto, 1954, p. 128

[= *Yokoyamoceras* Shimizu 1935, p. 198, *nom. nud.*]

Type species. *Holcodiscus kotoi* Jimbo, 1894, p. 33, pl. 21, fig. 2, by original designation.

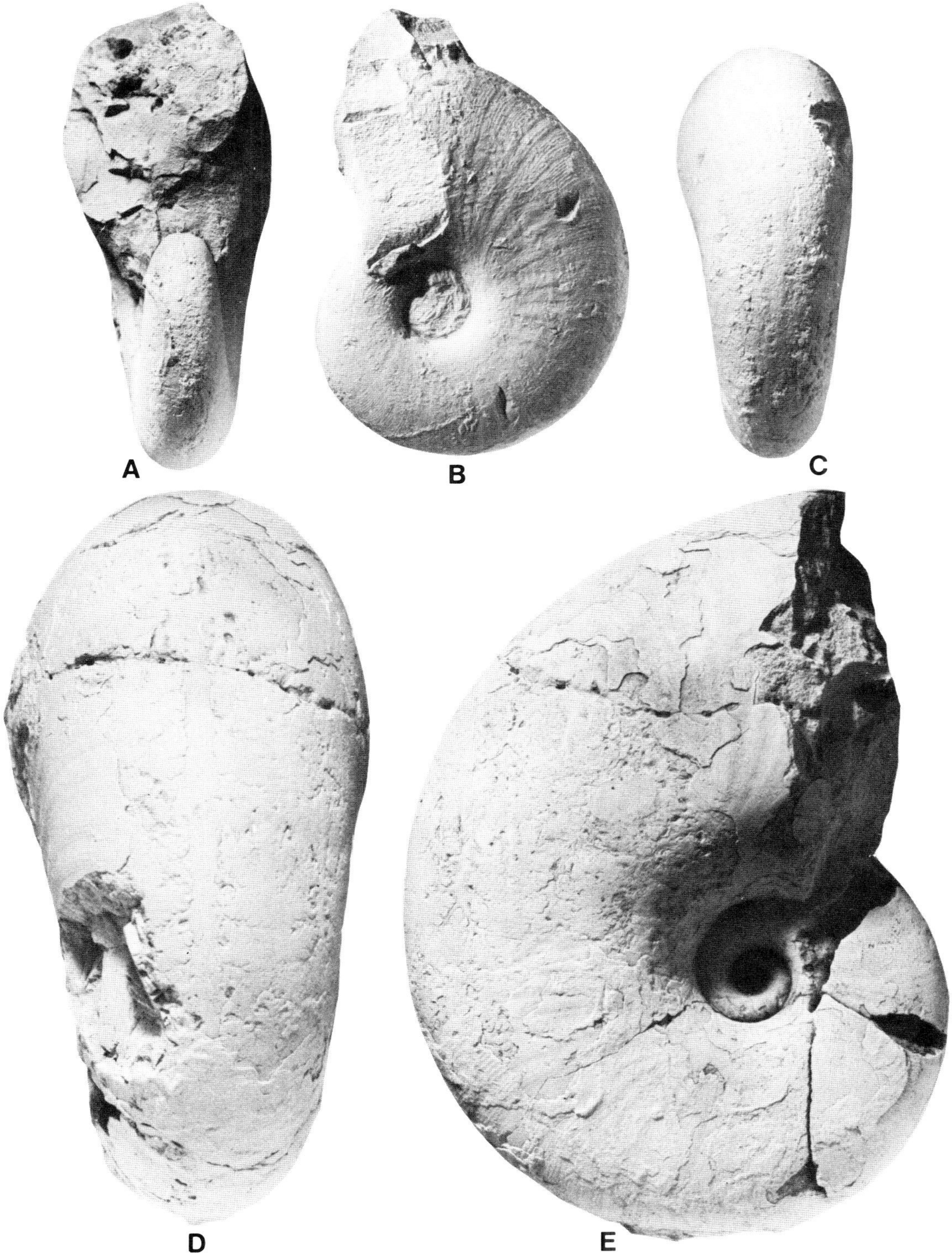

TEXT-FIG. 6. *Pseudophyllites indra* (Forbes, 1846). A–C, EMP unregistered, a syntype of *Gaudryceras colloti* de Grossouvre, 1894 (pl. 37, fig. 8), Upper Maastrichtian Calcaire à *Stegaster* between Gan and Rébénacq (Pyrénées-Atlantiques). D, E, the lectotype, BMNH C51068, the original of Forbes 1846, pl. 11, fig. 7, Valudayur Group, near Pondicherry, south India. All specimens ×1.

TEXT-FIG. 7. *Yokoyamoceras* species. A–E, *Y. jimboi* Matsumoto, 1955*a*; A–C, GK H5201; C, D, GT I-349, the holotype. F, G, *Y. kotoi* (Jimbo, 1894), GT I-107, the holotype. H, I, *Y. ornatum* Matsumoto, 1956, the holotype, GK H5210. J, K, *Y. minimum* Matsumoto, 1956, the holotype. GK H4204. All specimens ×2.

Discussion. Yokoyamoceras are a group of diminutive desmocerataceans characterized by the development of ventral tubercles in early maturity, following, and succeeded by a non-tuberculate stage. Ribs may link in pairs at the tubercles, while there are prominent constrictions. The type and other species bear lappets at maturity (text-fig. 7D; see also Matsumoto 1955*a*, text-figs. 12 and 13; pl. 9, fig. 4*a*; Obata *et al.* 1978, text-figs. 7 and 8; pl. 1, figs. 1–3), showing it to be a microconch genus. But of what? Wright and Matsumoto (1954), Matsumoto (1955*a*), and Wright (1957) all refer it to the Kossmaticeratinae, yet Matsumoto (1955*a*) drew attention to the resemblance between *Y. kotoi* and *Neopuzosia japonica* Spath, 1922*a* and between *Y. jimboi* Matsumoto, 1955*a* and *N. ishikawi* (Jimbo, 1894). Obata *et al.* (1978) subsequently illustrated a *Kitchinites* (*Neopuzosia*) *japonica* with lappets which precludes linking of *japonica* and *kotoi* as a dimorphic pair, while Matsumoto *et al.* (1972) note similarity between co-occurring lappeted *K.* (*N.*) *haboroensis* Matsumoto and Inoma, 1972 *in* Matsumoto *et al.* 1972 and *K.* (*N.*) *ishikawi* of larger size with normal aperture. I nevertheless suspect that at least some *Yokoyamoceras* are puzosiine microconchs, and note the striking development of lappets in microconch *Puzosia* Bayle, 1878 (Scholz 1978, text-fig. 2; Marcinowski 1980, pl. 6, fig. 11), *Mesopuzosia* Matsumoto, 1954 (Matsumoto *et al.* 1972, text-fig. 9), and *Hauericeras* de Grossouvre, 1894 (Obata *et al.* 1978, text-fig. 4; pl. 3) as supporting evidence. Unfortunately the single French specimen casts no light on the problem.

Yokoyamoceras aff. *jimboi* (Matsumoto, 1955*a*)

Plate 1, figs. 6–8; text-fig. 8E

Material. SP unregistered *ex* Arnaud Collection from Assise P¹, Verteillac (Dordogne).

Dimensions.

	D	*Wb*	*Wh*	*Wb*:*Wh*	*U*
SP unregistered	27·0(100)	9·1(33·7)	9·8(36·3)	0·93	7·1(26·3)

Description. The specimen is a wholly septate composite mould. Coiling moderately evolute; umbilicus shallow and of moderate width; umbilical wall low, flattened, subvertical. Whorl section compressed, whorl breadth to height ratio 0·93, with greatest breadth low on flank. Inner flanks broadly rounded, outer flanks flattened and convergent, venter arched, rounded at smallest diameter visible, broadening markedly as size increases. There are five prominent constrictions per whorl. These are markedly concave and prorsiradiate on the flanks, sweep forwards over the ventrolateral shoulder and cross the venter in a narrow convexity. The constrictions are preceded by a feeble collar-rib, most prominent over the venter, and succeeded by a much weaker rib, obvious only close to the umbilical shoulder. At the smallest diameter visible the flanks between constrictions are ornamented by very fine prorsiradiate concave ribs which more or less parallel constrictions. As size increases feeble bullae develop and can be seen to give rise to groups of ribs, which become feebly flexuous. Ventrolateral clavi appear on the last quarter whorl. There are three between the last constrictions; they link groups of two to three flank ribs, and are linked over the near-smooth venter by an effaced convex swelling.

Discussion. This remarkable specimen is the only *Yokoyamoceras* known outside the Japanese Islands. It most closely recalls *Y. jimboi* Matsumoto, 1955*a* (p. 153, pl. 9, figs. 4 and 5; text-fig. 13; 1956, p. 186, pl. 16, fig. 5; see text-fig. 7A–E) which has, however, five tubercles between constrictions and lacks the feeble bullae of the French form. *Y. kotoi* (Jimbo, 1894) (p. 33, pl. 21, fig. 2; Matsumoto 1955*a*, p. 151, text-figs. 11 and 12; see text-fig. 7F, G) develops feeble bullae at maturity, but has single ribs linking to ventral tubercles numbering up to twelve between constrictions and linked over the venter by narrow ribs. *Y. minimum* Matsumoto, 1956 (p. 185, pl. 16, fig. 6; Obata *et al.* 1978, p. 145, pl. 1, figs. 1 and 2; text-fig. 7; see text-fig. 7J, K) is closer, but ribs are near-obsolete on the inner flank with much coarser ventral clavi and a trace of a siphonal node on the adapical collar-rib. *Y.* (?) *mysticum* Matsumoto, 1955*a* (p. 155, pl. 10, fig. 1) has seven constrictions per whorl and straight rather than flexuous flank ribs. *Y. ornatum* Matsumoto, 1956 (p. 183, pl. 16, figs. 3 and 4) is much more coarsely and distantly ribbed (text-fig. 7H, I) and lacks tubercles.

Occurrence. As for material.

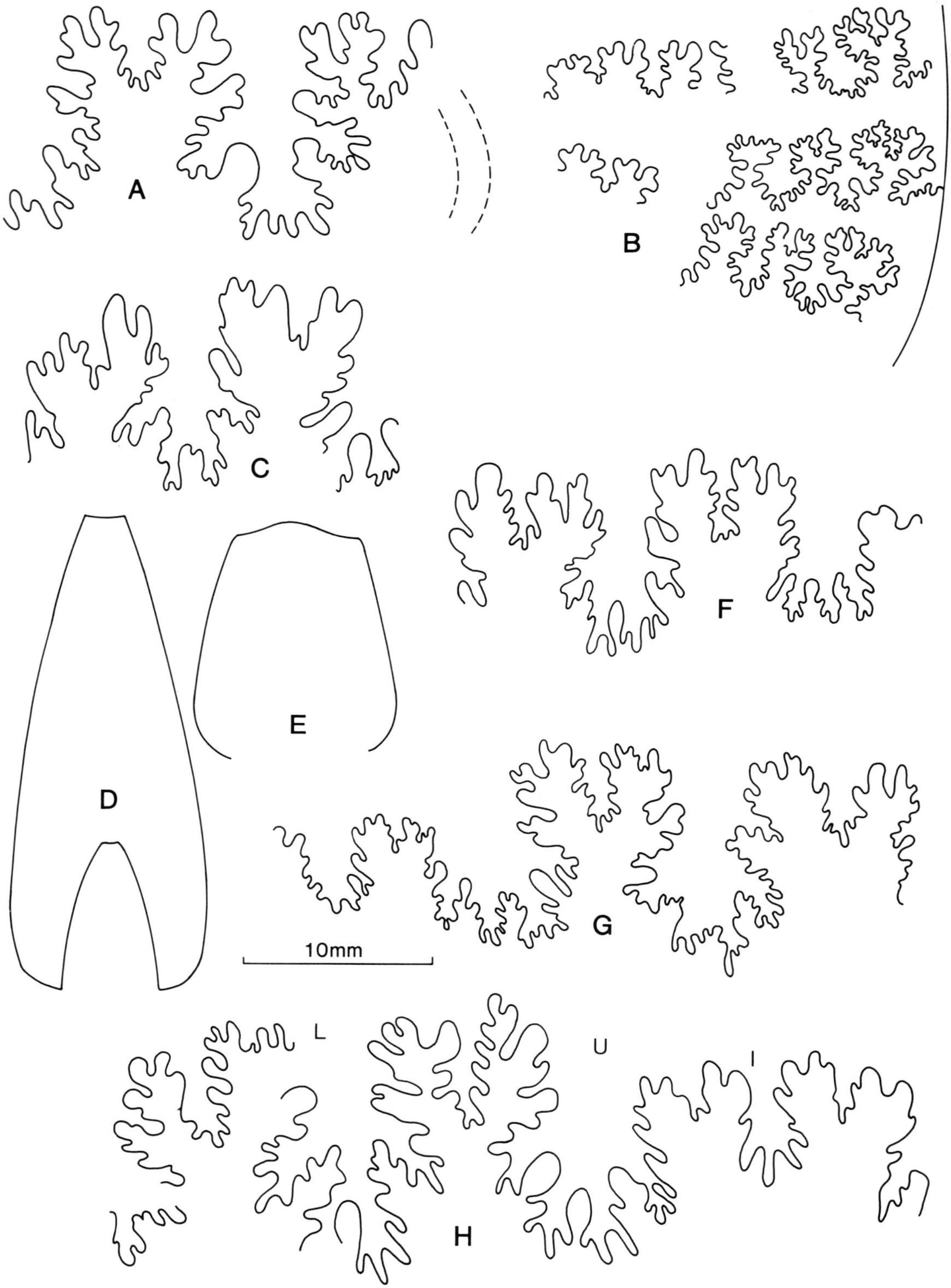

TEXT-FIG. 8. Suture lines and whorl sections of: A, C, *Baculites* sp. 1; B, D, *Placenticeras bidorsatum* (Roemer, 1841); E, *Yokoyamoceras* aff. *jimboi* (Matsumoto, 1955); F, G, *Baculites* sp. 2; H, *Baculites* sp. 4.

Subfamily HAUERICERATINAE Matsumoto, 1938, p. 193
Genus HAUERICERAS de Grossouvre, 1894, p. 219
[= *Schlueteria* Rollier, 1922, p. 359, *non* Fritsch *in* Fritsch and Kafka 1887, p. 33; *Pseudogardenia* Tomlin, 1930, p. 23; *Gardeniceras* Matsumoto and Obata, 1955, p. 134]

Type species. Ammonites gardeni Baily, 1855, p. 450, pl. 11, fig. 3, by original designation.

Hauericeras fayoli de Grossouvre, 1894

Text-fig. 9A–J

1894 *Hauericeras fayoli* de Grossouvre, p. 220 (*pars*), pl. 27, fig. 3.
non 1901 *Hauericeras fayoli* de Grossouvre; de Grossouvre, p. 382.
1925 *Hauericeras fayoli* de Grossouvre; Diener, p. 95.
1935 *Hauericeras gardeni* Baily; Brinkmann, p. 5.
?1961 *Hauericeras* sp. aff. *fayoli* de Gross; Collignon, p. 85, pl. 32, fig. 2.
1961 *Hauericeras fayoli* de Grossouvre; Collignon, p. 85.
1984 *Hauericeras fayoli* de Grossouvre; Kennedy and Summesberger, p 157, pl. 1, figs. 8 and 12; pl. 2, figs. 4, 13–15.

Type. Holotype, by monotypy, SP unregistered *ex* Arnaud Collection, the original of de Grossouvre 1894, pl. 27, fig. 3; refigured here as text-fig. 9A, B and said to be from Assise Q of Arnaud at Montmoreau (Charente); it is probably from Tercis (Landes) (see below).

Material. EMP unregistered. Without precise localization, the preservation of the specimen is that of Assise Q or R in northern Aquitaine.

Dimensions.

	D	*Wb*	*Wh*	*Wb*:*Wh*	*U*
EMP unregistered	62·0(100)	12·0(19·4)	19·0(30·6)	0·63	27·9(45·0)

Description. The specimen is a wholly septate internal mould. Coiling very evolute, the broad umbilicus shallow, comprising 45 % of the diameter. The whorl section is compressed (whorl breadth to height ratio 0·63), lanceolate, with a shallow impressed dorsal zone, sharp umbilical shoulder, flattened inner and convergent outer flanks (text-fig. 9C, F, G). The venter is narrowly arched on the internal mould, while it retains, in places, a spar-infilled keel (compare text-fig. 9G, I).

There are four constrictions per whorl. These are most prominent on the inner whorls (text-fig. 9E, F) and are effacing at the largest preserved diameter (text-fig. 9H). They are narrow, first appear on the umbilical wall, are straight and prorsiradiate on the inner flank, convex at mid-flank and concave on the outer flank with a strong projection over the venter, which they cross in a narrow, linguoid convexity.

The suture line includes moderately subdivided, narrow-stemmed bifid saddles and lobes. The umbilical lobe is strongly retracted.

Discussion. De Grossouvre (1894, p. 220, pl. 35, fig. 9) originally described the holotype of this species (text-fig. 9A, B) as being from Assise Q of Arnaud at Montmoreau. The preservation is clearly not that of this unit at the cited locality; instead, it is identical with that of the Campanian–Maastrichtian at Tercis (Landes). It is poorly preserved (text-fig. 9A, B), but has a comparable degree of involution, expansion rate, form and frequency of constrictions as the better preserved material described here. Specimens from the Silex de Mussidan mentioned by de Grossouvre (1894, p. 221) are preserved in the Arnaud Collection, labelled in de Grossouvre's hand. They are *Brahmaites brahma* (Forbes, 1846).

Hauericeras rembda (Forbes, 1846) (p. 111, pl. 7, fig. 3) from the Maastrichtian Valudayur Beds of southern India (lectotype is BMNH C51021) is a much more compressed species. The umbilicus is significantly smaller and the constrictions lack the striking sinuosity seen in the present species, although there is a distinct flexuosity in Stoliczka's larger fragment (1865, p. 63, pl. 33, fig. 5). The keel is also distinctive, with a remarkable hexagonal cross-section (ibid., fig. 5*a*).

H. sulcatum (Kner, 1848) (p. 8, pl. 1, fig. 3) is a further Maastrichtian species recently reviewed by Błaszkiewicz (1980, p. 41, pl. 54, fig. 3) and known from Poland, the USSR, and Bulgaria (see Tzankov

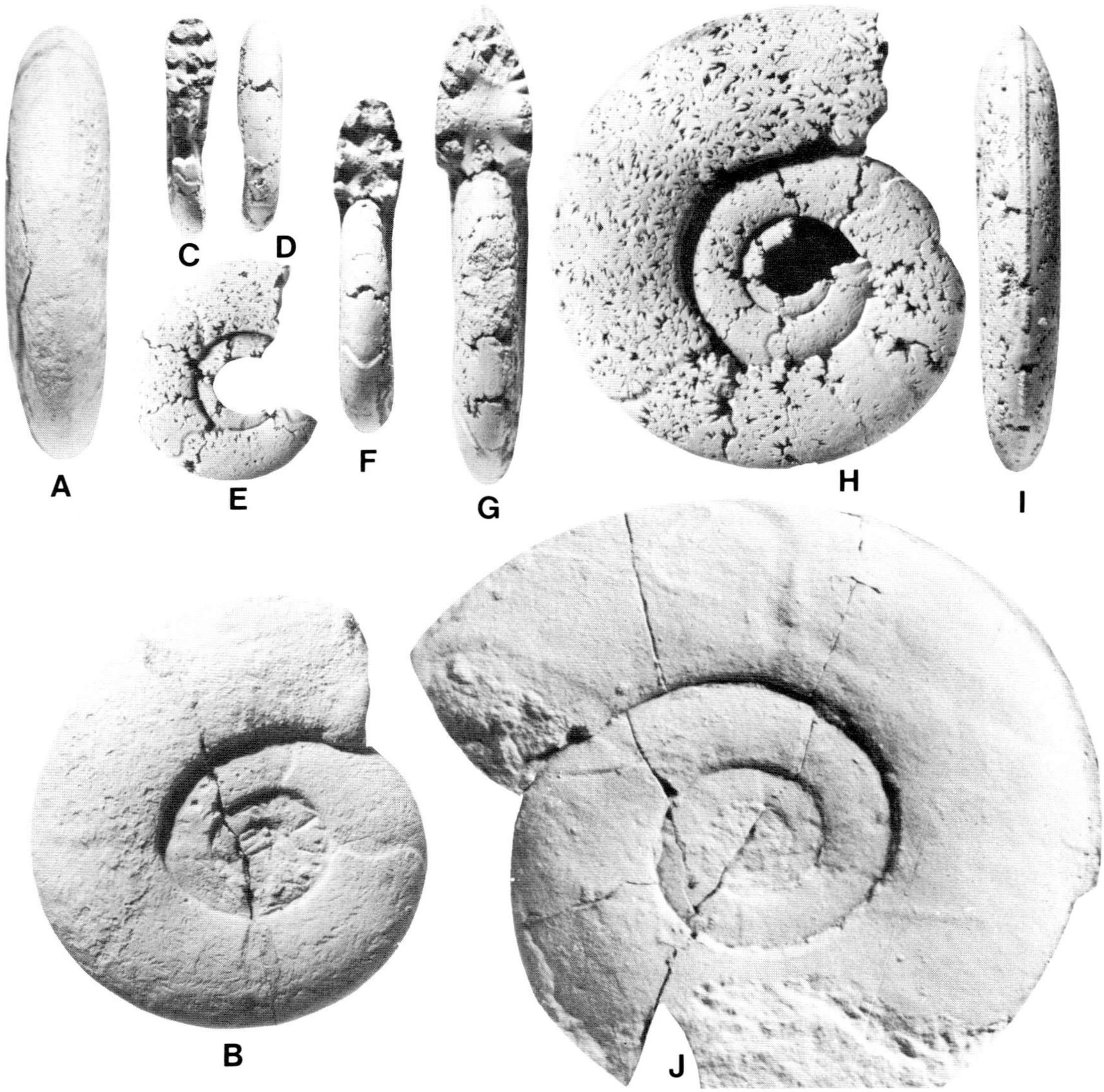

TEXT-FIG. 9. *Hauericeras fayoli* de Grossouvre, 1894. A, B are the holotype, SP unregistered *ex* Arnaud Collection, said to be from Assise Q at Montmoreau (Charente), but probably from Tercis (Landes); C–I, EMP unregistered, from its preservation from Assise Q or R in northern Aquitaine; J, specimen from the Upper Campanian Ultrahelvetic of the Gschliefgraben, Upper Austria, with keel and constrictions well preserved. Ulrich Roschger Collection, Gmunden/Vienna. All specimens × 1.

1964 for synonymy). It is more involute (U = 31%), higher-whorled, with more numerous constrictions (six to seven per whorl) that are prorsiradiate and concave rather than flexuous, as in the present form. *H. pseudogardeni* (Schlüter, 1871) (p. 54, pl. 16, figs. 3–6) and *H. welschi* de Grossouvre, 1894 (p. 222, pl. 35, fig. 9) are both much older species with, again, higher whorls, more numerous and less flexuous constrictions.

Occurrence. The precise horizon of the holotype is unknown but probably Upper Campanian–Lower Maastrichtian. The species also occurs in the Upper Campanian *polyplocum* Zone Ultrahelvetic of the Gschliefgraben, Upper Austria and there are doubtful records from the Lower Maastrichtian of Madagascar.

Hauericeras sp.

Material. SP unregistered *ex* Arnaud Collection from Assise P^3, Le Caillaud (Charente).

Discussion. This worn specimen shows the evolute coiling and compressed whorl section of *Hauericeras*. Although specifically indeterminate it has proportions recalling the Santonian–Campanian *H. gardeni* (Baily, 1855), known from the Gosau Beds of Austria as well as occurring widely in the Indo-Pacific region (see recent revision by Summesberger 1979, p. 133, pl. 6, fig. 27; text-fig. 19).

Occurrence. As for material.

Family KOSSMATICERATIDAE Spath, 1922*a*, p. 134
[*nom. transl.* Spath, 1923, p. 35 *ex* Kossmaticeratinae Spath, 1922*a*, p. 134]

Subfamily KOSSMATICERATINAE Spath, 1922*a*, p. 134
Genus BRAHMAITES Kossmat, 1897, p. 44 (151)

Type species. Ammonites brahma Forbes, 1846, p. 100, pl. 8, fig. 1, by original designation (Kossmat 1897, p. 44 (151)).

Discussion. Henderson and McNamara (1985, p. 67) review this genus and recognize two subgenera. *Brahmaites* (*Brahmaites*) has depressed whorls throughout ontogeny, umbilical bullae that disappear at around 70 mm, annular ribs that appear at 60 mm, infrequent constrictions preceded by prominent collars which bear siphonal tubercles in late growth. *B.* (*Anabrahmaites*) Yabe and Shimizu, 1924 (p. 79; type species, by original designation, *A. vishnu* Forbes, 1846, p. 100, pl. 7, fig. 9) (= *Subbrahmaites* Yabe and Shimizu, 1924, p. 79; type species *S. sachalinensis* Yabe and Shimizu, 1924, p. 77, pl. 13) has equidimensional to weakly compressed whorls bearing simple annular ribs, weak in youth but strong and widely spaced with siphonal tubercles at maturity, bullae being absent.

Brahmaites brahma (Forbes, 1846)

Text-fig. 10A–K

1846 *Ammonites Brahma* Forbes, p. 100, pl. 8, fig. 1.
1865 *Ammonites Brahma* Forbes; Stoliczka, p. 163, pl. 79, figs. 2–4.
1894 *Hauericeras fayoli* de Grossouvre, p. 220 (*pars*), *non* pl. 27, fig. 3.
1897 *Brahmaites Brahma* Forbes sp.; Kossmat, p. 45 (152) (*pars*), pl. 8 (19), fig. 9; ? *non* pl. 8 (19), figs. 7 and 8.
1901 *Hauericeras fayoli* de Grossouvre; de Grossouvre, p. 382.
?1907 *Brahmaites Brahma* Forbes; Boule *et al.*, p. 7 (27).
1925 *Brahmaites Brahma* Forbes; Diener, p. 103 (*pars*).
?1938 *Brahmaites Brahma* Forbes; Collignon, p. 97 (44), pl. 8, figs. 1 and 2.
1938 *Brahmaites brahma* Forbes; Roman, p. 390, pl. 40, fig. 377.
1957 *Brahmaites brahma* (Forbes); Wright, p. L376, fig. 491, 1.
?1971 *Brahmaites brahma* Forbes; Collignon, p. 20, pl. 647, figs. 2396 and 2397.

Types. Lectotype, here designated, BMNH C51029, the original of Forbes 1846, pl. 8, fig. 1*a*, *b* (text-fig. 10J, K). Paralectotype BMNH C51030 is the original of Forbes 1846, pl. 8, fig. 1*c*. Unfigured syntypes are BMNH C51031–51037, all *ex* Kaye and Cunliffe Collection, from the Lower Maastrichtian Valudayur Group of Pondicherri, south India.

Material. SP unregistered *ex* Arnaud Collection, five specimens from the Silex de Mussidan of Mussidan (Dordogne), Assise R^2 of Arnaud.

Description. The most complete specimen is an external mould of one flank of a specimen approximately 18 mm in diameter (text-fig. 10B, C). Coiling is evolute with a deep umbilicus. The umbilical wall slopes outwards, while flank and venter are broadly and evenly rounded, giving a uniform whorl section. There are four to five prominent narrow straight prorsiradiate collar-ribs per whorl, each associated with a parallel constriction. There are four to five primary ribs between constrictions. They arise at the umbilical seam, pass straight up the umbilical wall and strengthen into bullae at the umbilical shoulder. A prorsiradiate straight rib extends across the inner flank declining rapidly to leave a smooth venter.

Sutures not seen.

Discussion. Most specimens are smaller than the illustrated specimen. All match exactly the type and other material from Pondicherri in the British Museum (Natural History), as can be seen from text-fig. 10D–K, an ontogenetic series linking the French material with the much larger lectotype. Henderson and McNamara (1985, p. 70) have suggested that Kossmat's specimens are not *brahma*, referring the larger specimen from Kulmodu near Otacod to their new species, *B. kossmati* Henderson and McNamara, 1985 (p. 68, pl. 6, figs. 11–14; pl. 7, fig. 1; text-figs. 9*a*, *b*, and 10). This is an Upper Maastrichtian species (note that Kossmat's specimen is from a horizon above that of type *brahma*), more evolute, more depressed, and more densely ribbed than *brahma*. Kossmat's smaller specimen (1897, p. 45 (152), pl. 8 (19), fig. 8) has very fine dense ribs, arising in groups from bullae, resembling juvenile *Jacobites* (*Jacobites*) *nodulosus* (Marshall, 1926) (p. 171, pl. 21, fig. 4; pl. 36, figs. 1–3) as represented by *B. rotundus* Marshall, 1926 (p. 173, pl. 21, fig. 9; pl. 31, figs. 4 and 5) a synonym according to Henderson (1970, p. 41).

Madagascan specimens referred to *B. brahma* by Collignon (1938, pl. 8, figs. 1 and 2; 1971, pl. 647, figs. 2396 and 2397) have coarser ornament and different whorl proportions. Specimens figured by Collignon (1971) come from localities yielding *Anapachydiscus fresvillensis* (Seunes, 1890) (e.g. 1971, p. 30, pl. 652, fig. 2408) an Upper Maastrichtian species (*fide* Kennedy 1986*a*, p. 46) and may thus be younger than the types of *brahma*.

Occurrence. Lower Maastrichtian of Pondicherri, south India, Silex de Mussidan, Assise R^2 of Arnaud at Mussidan (Dordogne). Allied forms occur in the Maastrichtian in Madagascar.

Family PACHYDISCIDAE Spath, 1922*a*, p. 132
[*nom. transl.* Spath, 1923, p. 39; *ex* Pachydiscinae Spath, 1922*a*]

Discussion. The Pachydiscidae as currently conceived includes two groups of genera. Generally large forms with ribs and bullae only or near-smooth are genera *Eopachydiscus* Wright, 1955, *Lewesiceras* Spath, 1939, *Tongoboryceras* Houša, 1967, *Nowakites* Spath, 1922*a*, *Canadoceras* Spath, 1922*a*, *Patagiosites* Spath, 1953, *Anapachydiscus* Yabe and Shimizu, 1926, *Eupachydiscus* Spath, 1922*a*, *Pachydiscus* Zittel, 1884, and *P.* (*Neodesmoceras*) Matsumoto, 1947; generally small forms developing ventral or ventrolateral tubercles at maturity are *Menabonites* Houša, 1967, *Pseudojacobites* Spath, 1922*a*, *Tuberodiscoides* Collignon, 1966, *Teshioites* Matsumoto, 1955*b*, *Menuites* Spath, 1922*a*, *Urakawites* Matsumoto, 1955*b*, and *Pseudomenuites* Matsumoto, 1955*b*. *Pachydiscoides* Spath, 1922*a*, is known from juveniles only, while *Bayleites* Collignon, 1952 is a Jurassic perisphinctacean, as discussed below.

Spath (1922*a*, p. 123) regarded *Menuites* as an offshoot of *Parapachydiscus*, while Matsumoto (1955*b*) regarded it as an offshoot of *Anapachydiscus*, noting that juveniles before the bituberculate stage were 'hardly distinguishable from immature shells of *Anapachydiscus*' (p. 156). *Urakawites* were

---

TEXT-FIG. 10. *Brahmaites brahma* (Forbes, 1846). A–C, SP unregistered *ex* Arnaud Collection, Silex de Mussidan of Mussidan (Dordogne); D–K, ontogenetic series linking French and Indian specimens. D is BMNH C3564; E is BMNH C51033, a paralectotype; F, H, BMNH 24195; G, I, BMNH C51032, a paralectotype; J, K, the lectotype, BMNH C51029, the original of Forbes 1846, pl. 8, fig. 1. All from the Lower Maastrichtian Valudayar Beds of Pondicherry, south India. All specimens ×1.

A
B
C
D
E
F
G
H
I
J
K

said to be offshoots of *Eupachydiscus* and certain *Anapachydiscus*, *Pseudomenuites* offshoots of *Pachydiscus* of the *neubergicus–gollevillensis* group, and *Teshioites* close allies of *Canadoceras*. In his 1955*b* paper Matsumoto regarded the small bituberculate genera as neotenous derivatives of 'normal' pachydiscids, and specifically rejected the idea that they might be sexual dimorphs (p. 155). In his latest description of bituberculate pachydiscids (1984*b*) he regards dimorphism as an explanation of the occurrence of these two disparate groups of ammonites as no more than a possibility. My own observations and those of Kennedy and Summesberger (1984) suggest that in some if not all cases, pachydiscids are dimorphic. This presents taxonomic problems, since although some species of microconch genera can be linked to some species of macroconch genera the type species cannot always be so linked, while individual genera of microconchs in some cases correspond to several macroconch genera. A conservative approach is best maintained in those cases where this uncertainty exists, but I note the following correlations:

1. *Eopachydiscus*: dimorphism not recognized, probably reflected in adult size differences alone.

2. *Lewesiceras*: in *L. mantelli* at least, there is striking size dimorphism in English Chalk Rock material. Marcinowski (1972, p. 532, pl. 4, figs. 1, 2) has shown similar size dimorphism in what he calls *L. peramplum* Mantell, 1822) from Poland.

3. *Canadoceras*: *Teshioites ryugasensis* Matsumoto, 1955*b* (p. 174, pl. 36, figs. 1 and 2; pl. 37, fig. 3; text-fig. 8) is clearly the microconch of *C. kossmati*. *T. teshioensis* Matsumoto, 1955*a* (p. 177, pl. 37, fig. 2; pl. 35, fig. 3) is clearly the microconch of *C. yokoyamai* (Jimbo, 1894) (p. 177, pl. 18, fig. 3). The microconch of the type species, *C. newberryanum* Meek, 1876*a* (p. 47) has not been recognized.

4. *Nowakites*: material from the Corbières (Aude) shows typical ribbed specimens occurring with smaller bituberculate ones, figured as *Sonneratia savini* de Grossouvre, 1894 (p. 152, pl. 37, fig. 4) and referred to *Nowakites* by Spath (1922*a*, p. 124) and to *Teshioites* n. sp. (?) by Matsumoto, 1955*b* (p. 174).

5. *Anapachydiscus*: Matsumoto (1984*b*) has described co-occurring *A. fascicostatus* (Yabe, 1921) (the type species) and *Menuites sandai* Matsumoto, 1984*b*, a clear dimorphic pair. *A. wittekindi* (Schlüter, 1876) co-occurs with *Urakawites auritocostatus* (Schlüter, 1872*a*). The lectotype of the latter (*Ammonites proteus* Schlüter, 1867, pl. 3, fig. 2, designated by Matsumoto 1955*b*, p. 169) seems very different from *wittekindi*, but this is due to crushing. Study of large topotype collections in the Museum für Naturkunde, Berlin and the Institut für Geologie und Paläontologie, Bonn show *wittekindi* and *auritocostatus* to have identical early whorls. Also occurring in the Haldem fauna is the lectotype of *Pachydiscus ambiguus* de Grossouvre, 1894. Previous authors have interpreted this in terms of the Aquitaine specimen figured by de Grossouvre (pl. 29, fig. 3), designated lectotype by Matsumoto (1955*b*, p. 169, footnote). In fact, Diener (1925, p. 112) designated the second specimen mentioned by de Grossouvre, the original of Schlüter (1872*a*, pl. 22, figs. 6 and 7) 'type' (a valid lectotype designation), and *ambiguus*, the type species of *Pseudomenuites* is no more than a crushed microconch body-chamber of *Ammonites haldemsis* Schlüter, 1867 (compare text-fig. 11A–D and F, G). Uncrushed specimens from the Campanian White Limestone of northern Ireland show *Anapachydiscus* associated with *Menuites portlocki* (Sharpe, 1855) in a dimorphic pair while a similar association is seen in Poland (Błaszkiewicz 1980). Quite which of the above names has priority is outside the scope of this review. In the Campanian of British Columbia a further pair is *Eupachydiscus perplicatus* (Whiteaves, 1903) and *Pachydiscus binodatus* (Whiteaves, 1903), the latter referred to *Urakawaites* by Matsumoto (1955*b*, p. 179).

6. *Eupachydiscus*: Immel *et al.* (1982) described co-occurring *E. isculensis* (Redtenbacher, 1873) and a small bituberculate form, *Pseudomenuites katschthaleri* Immel *et al.*, 1982, again interpreted as a dimorphic pair.

7. *Pachydiscus* (*Pachydiscus*): Kennedy and Summesberger (1984) show *P.* (*P.*) *koeneni* de Grossouvre, 1894 and the diminutive bituberculate *Ammonites haldemsis* Schlüter, 1867, the latter referred to *Pseudomenuites* by Matsumoto (1955*b*) to be a dimorphic pair. The lectotype and paralectotype of *P. ambiguus* de Grossouvre, 1894 are also microconchs of the *Pachydiscus* (P.) *oldhami* (Sharpe, 1855)–*koeneni* group, as discussed further below. Unpublished work with Herbert Klinger on the Zululand faunas suggests an association of *Menuites*-like forms with *Pachydiscus*

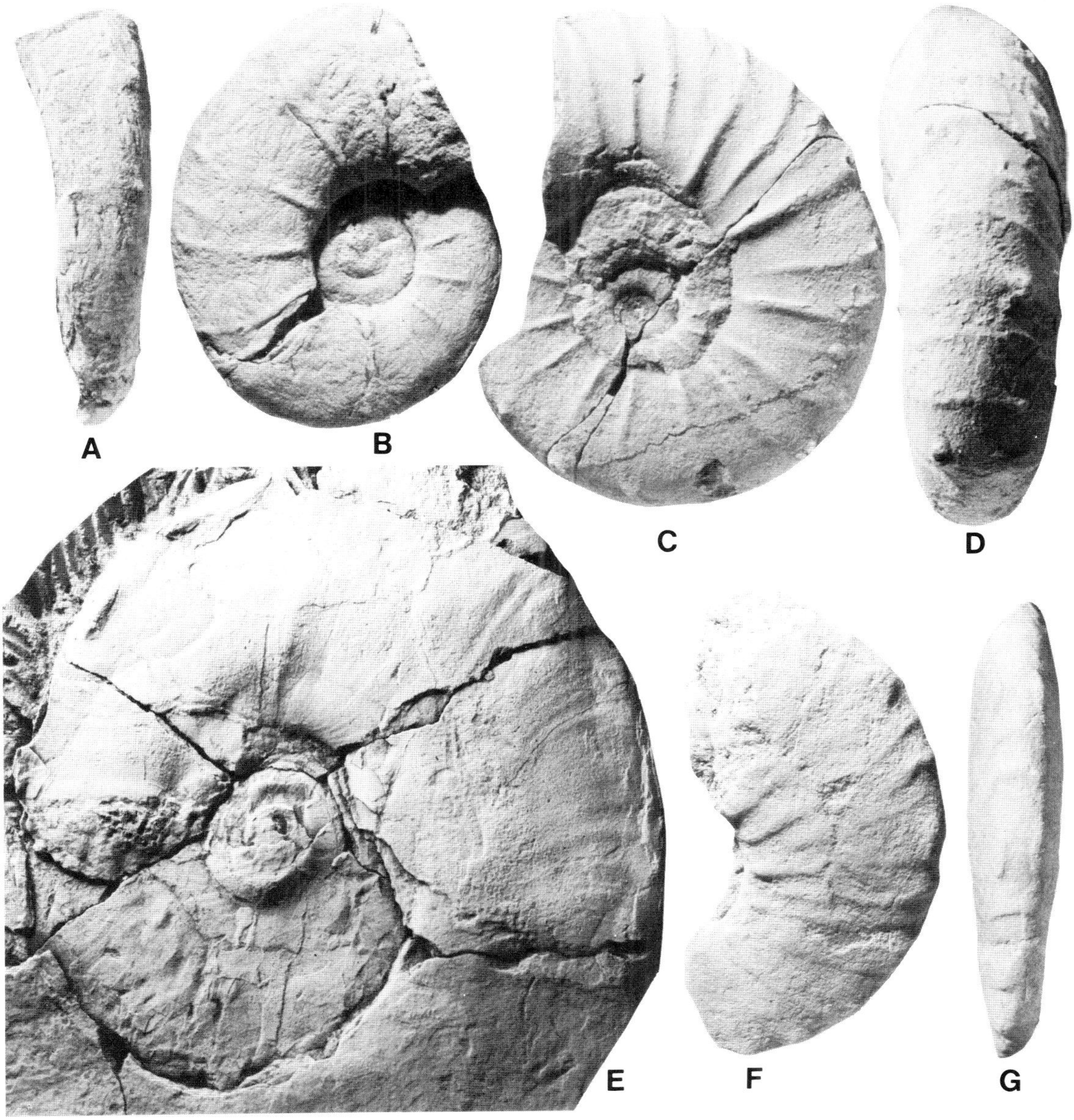

TEXT-FIG. 11. A–D, F, G, *Pachydiscus* (*Pachydiscus*) *haldemsis* (Schlüter, 1867). A, B, the lectotype, a microconch, GPIB unregistered, the original of Schlüter 1867, pl. 3, fig. 1, Upper Campanian of Haldem, Westphalia; C, D, uncrushed adult microconch, Upper Campanian Ultrahelvetic of the Gschliefgraben, Upper Austria, in the Ulrich Roschger Collection, Gmunden/Vienna; F, G, the lectotype of *P.* (*P.*) *ambiguus* de Grossouvre, 1894, the original of Schlüter 1872*a*, pl. 22, figs. 6 and 7, from the same horizon and locality as A and B, GPIB 5*b*. E, *Placenticeras milleri* (Hauer, 1866), the best preserved syntype in the GBA Collections, Steinbruch, Steiermark, Austria. All specimens × 1.

(*Neodesmoceras*). As yet dimorphism has not been recognized in the type species *P.* (*P.*) *neubergicus* (Hauer, 1858).

8. *Menuites*: this microconch genus includes species that correspond to *Anapachydiscus* as noted above. The type species, *M. menu* (Forbes, 1846), occurs with juvenile pachydiscids only in south India, and I have not recognized the corresponding dimorph. Matsumoto (1955*b*) notes similarity to *Anapachydiscus fascicostatus*.

Genus PACHYDISCUS Zittel, 1884, p. 466

Type species. Ammonites neubergicus Hauer, 1858, p. 12, pl. 2, figs. 1–4, by the subsequent designation of de Grossouvre (1894, p. 177).

Diagnosis. Compressed to depressed, moderately involute, with high, flat, or convex sides. Ornamented by straight to feebly curved primary ribs, feebly bullate or not, with or without well-differentiated secondaries. Ornament may persist, reduce to umbilical ribs, with or without feeble bullae or disappear completely at maturity. Macroconchs may reach a large size. Microconchs, known in a few species only, are very small with ventrolateral tubercles.

Occurrence. Campanian to Maastrichtian, world-wide.

Subgenus PACHYDISCUS Zittel, 1884, p. 466

[= *Parapachydiscus* Hyatt, 1900, p. 570; *Joaquinites* Anderson, 1958, p. 218; *Pseudomenuites* Matsumoto, 1955*b*, p. 169]

Diagnosis. Pachydiscus in which ornament persists to middle growth, with umbilical or lateral ribs retained to maturity.

Discussion. P. (*Pachydiscus*) *neubergicus* (Hauer, 1858) and species such as *P. egertoni* (Forbes, 1846), *P. jacquoti* (Seunes, 1890), *P. gollevillensis* (d'Orbigny, 1850) (type species of *Parapachydiscus*), *Pachydiscus compressus* (Spath, 1922*a*) *P. llareni* Wiedmann, 1960, *P. carinthiacus* Thiedig and Wiedmann, 1976, *P. subcompressus* Matsumoto, 1954 (of which *P. fascicostatum* (Anderson, 1958), the type species of *Joaquinites* is a synonym), and *P. excelsus* Matsumoto, 1979 form a compact, predominantly Maastrichtian group probably derived paedomorphically from the Upper Campanian *P. perfidus* de Grossouvre, 1894 (p. 213, pl. 34, fig. 1; Kennedy and Summesberger 1984)–*P. oldhami* (Sharpe, 1855) (p. 32, pl. 14, fig. 2) group. The Campanian and some Maastrichtian species differ in being more densely and complexly ribbed in many cases and are often inflated, with massive whorls. This group, long-recognized (Matsumoto 1959*a*, p. 41) first appears in Europe with *P. duelmensis* (Schlüter, 1872*a*) in the Lower Campanian and there is a series of species in the higher Campanian including *P. launayi* de Grossouvre, 1894 and *P. colligatus* (Binkhorst, 1861). Maastrichtian representatives include *P. epiplectus* (Redtenbacher, 1873).

Campanian species that are compressed like the *neubergicus* group, but have complex, dense ribbing when young, include *P. oldhami* (Sharpe, 1855) and *P. koeneni* de Grossouvre, 1894. The latter is the only species referred to *P.* (*Pachydiscus*) in which dimorphism has been recognized, with the large, ribbed *koeneni* the macroconch, and the small, tuberculate *haldemsis* Schlüter, 1867 the microconch (Kennedy and Summesberger 1984, p. 158, pl. 4, figs. 1–5; pl. 5, fig. 1; pl. 6, fig. 2; pl. 7, figs. 1–11; pl. 13, fig. 1), the latter name having priority. In the case of *P.* (*P.*) *neubergicus* I have so far failed to recognize dimorphism of the type seen in *haldemsis* (= *koeneni*) and it may be that this will provide a basis for generic or subgeneric separation within the presently broadly conceived subgenus.

P. (*Neodesmoceras*) Matsumoto, 1947, p. 39 (republished in English 1951, p. 24 = *Neodesmoceras* Matsumoto, 1938, p. 193 (*nom. nud.*)) includes the type species, *P.* (*N.*) *japonicus* Matsumoto, 1947, *P.* (*N.*) *mokotibense* Collignon, 1952, *P.* (*N.*) *obsoletiformis* Jones, 1963, and *P.* (*N.*) *gracilis* Matsumoto, 1979. These are exclusively Maastrichtian and they are large and nearly smooth at maturity. Unpublished work by Herbert Klinger and myself shows that in *P.* (*N.*) *mokotibense* at least, the inner whorls have very distant primary ribs when young, while giant discs, still septate at 350 mm are macroconchs, associated with *Menuites*-like tuberculate microconchs adult at 110 mm diameter. Lower Campanian *Pachydiscus* of *duelmensis* type probably arose paedomorphically from forms such as *Eupachydiscus jeani* (de Grossouvre, 1894) (p. 127, pl. 26, fig. 5) of the Santonian, the inner whorls of which are finely and densely ribbed.

Occurrence. Lower Campanian to Upper Maastrichtian. World-wide.

Pachydiscus (*Pachydiscus*) *duelmensis* (Schlüter, 1872*a*)

Text-fig. 12

1872*a* *Ammonites Dülmensis* Schlüter, p. 52, pl. 16, figs. 1 and 2.
1894 *Pachydiscus dülmensis* Schlüter sp.; de Grossouvre, p. 199, pl. 20.
1925 *Pachydiscus Dülmensis* Schlüter; Diener, p. 105.
1952 *Pachydiscus dülmensis* Schlüter; Collignon, p. 90.
1955 *Pachydiscus dülmensis* Schlüter; Collignon, p. 81.

Discussion. The only Aquitaine specimens of this species were collected by Arnaud from Assize P^1 at Le Maine Bardon, and Chardurie in the Commune of Charmant (Charente), one of which was figured

TEXT-FIG. 12. *Pachydiscus* (*Pachydiscus*) *duelmensis* (Schlüter 1872*a*), the figured syntype GPIB 47, the original of Schlüter 1872*a*, pl. 15, figs. 1 and 2, Lower Campanian of Dülmen, Westphalia, German Federal Republic, × 0·7.

by de Grossouvre (1894, pl. 20). Schlüter's figured specimen is shown in text-fig. 12. It is identical to the French specimen, which I have not traced. This is a highly distinctive species, easily separated from all other *Pachydiscus* (*Pachydiscus*) found in Aquitaine by its massive depressed whorls and very fine, dense, wiry ribbing.

Occurrence. Lower Campanian: Assize P^1 in northern Aquitaine, the Vaals Formation in Belgium and The Netherlands, and Dülmen, Westphalia.

Pachydiscus (*Pachydiscus*) *colligatus* (Binkhorst, 1861)

Text-figs. 13 and 14

1861	*Ammonites colligatus*, Nobis; Binkhorst, p. 25 (*pars*), pl. 8, fig. 1 only.
?1891	*Pachydiscus fresvillensis* Seunes; Seunes, p. 14, pl. 12 (3), fig. 1.
1894	*Pachydiscus colligatus* Van Binkhorst sp. emend. A. de Grossouvre; de Grossouvre, p. 202 (*pars*), pl. 33, *non* pl. 24, figs. 1 and 3 (= *Anapachydiscus fresvillensis* (Seunes, 1890)).
1894	*Pachydiscus van den broecki* de Grossouvre, p. 207.
?1910	*Pachydiscus colligatus* Binkhorst, 1861; Pruvost, p. 368.
1938	*Parapachydiscus colligatus* von Binkhorst; Collignon, p. 80 (30), pl. 5, fig. 1.
In press *b*	*Pachydiscus colligatus* (Binkhorst, 1861); Kennedy, pl. 1, figs. 1 and 2; pl. 2, figs. 1 and 2; pl. 3; pl. 4, figs. 4 and 5 (with full synonymy).

Types. Lectotype, here designated is the original of Binkhorst 1861, pl. 8. The specimen is in the Collections of the Museum für Naturkunde, Berlin, and is from the Upper Campanian of Jauche (Brabant), Belgium. The surviving paralectotypes belong to several other species: *Pachydiscus* (*Pachydiscus*) *jacquoti* Seunes, 1890, *Anapachydiscus fresvillensis* (Seunes, 1890), *Eupachydiscus* cf. *levyi* (de Grossouvre, 1894) as discussed by Kennedy (in press *b*).

Material. SP unregistered *ex* Toucas Collection, labelled 'Labinde, rive gauche sur les bords de la Dordogne, Maestrichtien'. M. Neumann Collection (Paris): a specimen from benthic foraminifer Biozone CVI, equivalent to Assise Q of Arnaud, Journiac-Le Dognon, north of Le Buge (Dordogne).

Dimensions.

	D	*Wb*	*Wh*	*Wb*:*Wh*	*U*
De Grossouvre 1894, pl. 33 (taken from photograph)	172(100)	5·5(398)	80.4(46·5)	0·86	40(73·3)
Lectotype	260(100)	90 (240)	101 (45·3)	0·89	53(24)

Description. Coiling involute with umbilicus comprising 23% of diameter; of moderate depth, with a rounded wall. Whorl section depressed (whorl breadth to height ratio less than 0·9) with greatest breadth just outside umbilical shoulder. At the smallest diameter seen, ornament consists of well-developed umbilical bullae that give rise to pairs of blunt, concave prorsiratiate ribs. These are narrower than the interspaces, and sweep forwards over the ventrolateral shoulder to cross the venter in a marked convexity. Occasional intercalated ribs arise low on the flank. In middle growth the umbilical bullae decline, the number of intercalatories increases, and the inner flank ornament progressively declines. The species appears to lose all but feeble blunt ventral ribbing at maturity.

Discussion. As reviewed elsewhere (Kennedy in press *b*) this species has suffered a complex history of misinterpretation and usage. Commonly regarded as a Maastrichtian index, I have shown that the lectotype is actually datable as Campanian from microfossils in its matrix. Most specimens referred to it are *A. fresvillensis* (Seunes, 1890) of the Upper Maastrichtian, but the lectotype (text-fig. 13) is quite distinct and can be linked to the fine specimen from Assise Q at Royan (Charente-Maritime) figured by de Grossouvre (1894, pl. 33). I have been unable to trace this specimen, but other, poorer material noted above belongs to the same species (text-fig. 14). It would be inappropriate to repeat my long discussion of the species (Kennedy in press *b*) here. Suffice it to say that when compared with other similar species known from northern Aquitaine, *P.* (*P.*) *duelmensis* (Schlüter, 1872*a*) (text-fig. 12) has a much more massive shell with a depressed whorl section and very fine, wiry ribs. *P.* (*P.*) *launayi* (de Grossouvre, 1894) (p. 38; Pl. 2, figs. 1 and 2; Pl. 7, figs. 6 and 7; Pl. 10, fig. 15; Pl. 13, figs. 2, 3, 6)

A

B

TEXT-FIG. 13. *Pachydiscus* (*Pachydiscus*) *colligatus* (Binkhorst, 1861), the lectotype, the original of Binkhorst 1861, pl. 8, MNB unregistered, Upper Campanian of Jauche, Belgium. A, detail of ventral ornament, × 1. B, side view, × 0·65.

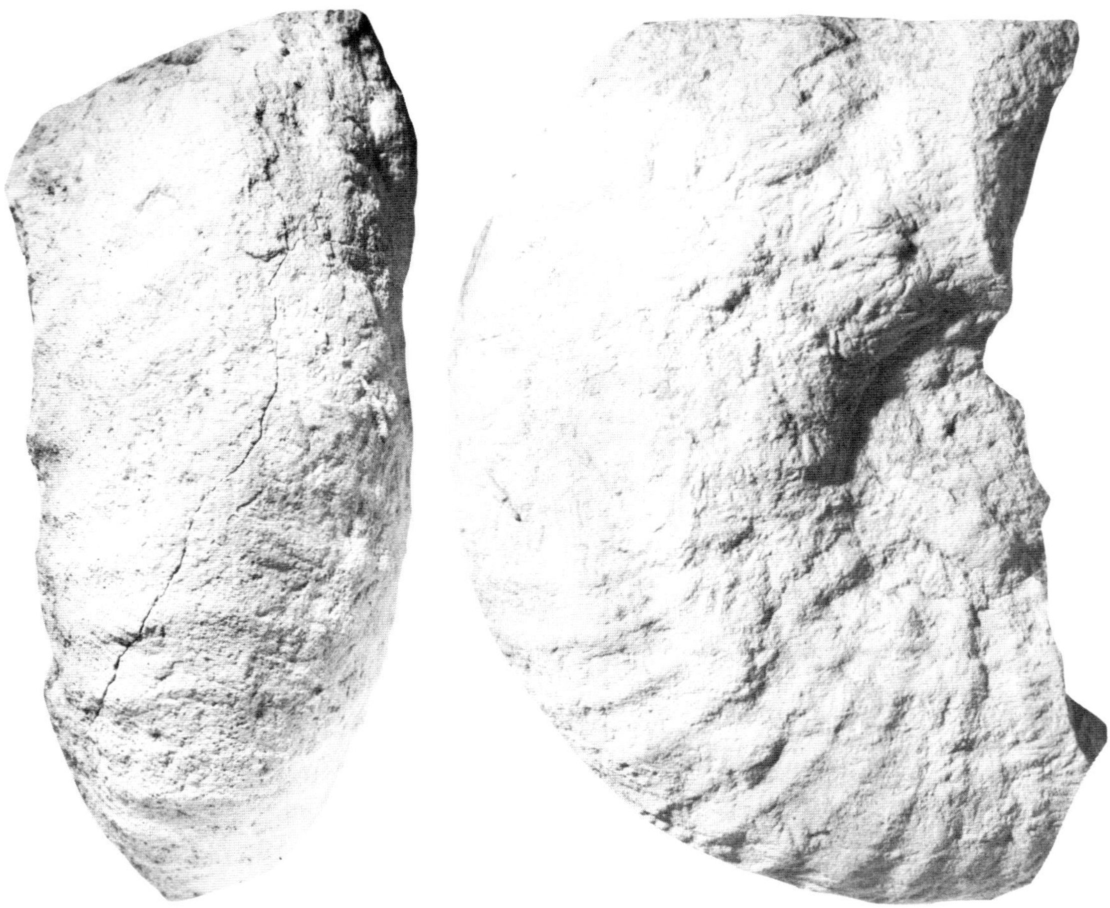

TEXT-FIG. 14. *Pachydiscus* (*Pachydiscus*) *colligatus* (Binkhorst, 1861). SP unregistered *ex* Toucas Collection, labelled 'Labinde, rive gauche sur les bords de la Dordogne', × 1.

is, in middle growth, depressed, generally more evolute, with long primaries that are straight on the flanks, arise singly at the umbilical shoulder, and are separated by single short intercalatories.

Occurrence. Upper Campanian, Assise P^3(?) and Q in northern Aquitaine, Upper Campanian of Brabant, Belgium, Madagascar, and possibly, Tercis (Landes).

Pachydiscus (*Pachydiscus*) *launayi* (de Grossouvre, 1894)

Plate 2, figs. 1 and 2; Plate 7, figs. 6 and 7; Plate 10, fig. 15; Plate 13, figs. 2, 3, 6; text-figs. 4C and 5B.

1894 *Pachydiscus launayi* de Grossouvre, p. 184, pl. 19.
1938 *Pachydiscus* aff. *launayi* de Grossouvre; Collignon, p. 10, pl. 1, fig. 2.
1952 *Eupachydiscus Launayi* de Grossouvre; Collignon, p. 35, pl. 5, fig. 1; text-fig. 6.
1952 *Eupachydiscus* aff. *Launayi* de Grossouvre; Collignon, p. 38, pl. 6, fig. 2.
1955 *Eupachydiscus Launayi* de Grossouvre; Collignon, p. 36, pl. 5, fig. 1; text-fig. 6.
1955 *Eupachydiscus* aff. *Launayi* de Grossouvre; Collignon, p. 39, pl. 6, fig. 2.
1969 *Eupachydiscus Launayi* Gross.; Collignon, p. 64, pl. 538, fig. 2112.
1984 *Pachydiscus launayi* de Grossouvre; Bilotte, pl. 37, fig. 1.

Holotype. By monotypy, SP unregistered *ex* Arnaud Collection, the original of de Grossouvre 1894, pl. 19, from the lower part of Assise P[1], Voulgézac (Charente).

Material. SP unregistered *ex* Arnaud Collection, three specimens: two from Assise P[1] at Maine-Bardon (Charente); one from Assise P[2], St Hilaire-de-Montmoreau (Charente).

Dimensions.

	D	*Wb*	*Wh*	*Wb*:*Wh*	*U*
Holotype	155·0(100)	— (—)	66·5(43·0)	—	46·0(29·7)
at:	113·0(100)	53·5(47·3)	51·0(45·1)	1·05	30·3(26·8)
SP unregistered	125·0(100)	58·5(46·8)	52·5(42·0)	1·1	37·5(30·0)
SP unregistered	67·5(100)	— (—)	30·5(45·2)	—	15·5(23·0)

Description. Coiling moderately evolute, with 56% of the previous whorl covered. Umbilicus comprises 23–30% of diameter, increasing through ontogeny, is of moderate depth with a rounded, outwards-inclined wall. Whorl section slightly depressed, reniform, with greatest breadth just outside umbilical shoulder. On the phragmocone during middle growth stages twenty primary ribs per whorl arise at the umbilical shoulder without developing bullae. They are narrow, distant, rounded, and sharply demarcated from the interspaces.

They are straight and prorsiradiate across the flank, then flex forwards and are convex on the ventrolateral shoulders, crossing the venter in a broad convexity. In a few cases ribs arise in pairs, while there are short intercalatories inserted on the flank to give a total of forty-two ribs per whorl. The secondaries decline on the body-chamber and ornament consists of slightly flexuous prorsiradiate primaries only.

Suture deeply incised, typical for genus (Pl. 7, fig. 7).

Discussion. All specimens are rather worn, notably the holotype. The side view given by de Grossouvre is reversed and retouched. The present form closely resembles species generally referred to *Pachydiscus*, lacking the coarse ribbing of *Eupachydiscus*, to which it is referred by some authors. Of comparable species recorded from northern Aquitaine, *P.* (*P.*) *duelmensis* (Schlüter, 1872*a*) (p. 52, pl. 16, figs. 1 and 2) is easily distinguished by its very fine, dense, crowded ribs, numbering in excess of sixty per whorl (text-fig. 12). *P.* (*P.*) *colligatus* (Binkhorst, 1861) (see p. 36; text-fig. 14) has a compressed whorl section with a much smaller umbilicus during middle and later growth (text-fig. 13). The rib density is higher, the ribs curved and prorsiradiate, with, in some specimens, pairs arising from distinct bullae.

Occurrence. Lower Campanian; Assises P[1] and P[2] in northern Aquitaine. Lower Campanian of Madagascar.

Pachydiscus (*Pachydiscus*) cf. *complanatus* Collignon, 1952

Plate 6, figs. 6–8

1952 *Pachydiscus complanatus* Collignon, p. 77, pl. 27, fig. 1.
1955 *Pachydiscus complanatus* Collignon; Collignon, p. 72, pl. 27, fig. 1.
1969 *Pachydiscus complanatus* Collignon; Collignon, p. 64, pl. 538, fig. 2113.

Types. Holotype, by original designation, Collignon's specimen no. 362, now in the École des Mines Collections, currently housed in the Université Claude Bernard, Lyon. There is one paratype. Both are from the Lower Campanian of Menabe, Madagascar.

Material. SP unregistered *ex* Arnaud Collection, from the Assise P[2], St Laurent-de-Belzagot (Charente).

Dimensions.

D	*Wb*	*Wh*	*Wb*:*Wh*	*U*
102·5(100)	30·0(29·3)	48·0(46·8)	0·63	22·5(22·0)

Description. The specimen is a crushed composite mould. Coiling involute, with a small umbilicus comprising 22% of diameter. The umbilicus is of moderate depth with a rounded wall. Whorl section compressed oval (whorl breadth to height ratio 0·68 on the least deformed section) with flattened convergent flanks and rounded venter. The mould is smooth except for a trace of distant ventral ribs in places.

Discussion. This remarkable, near-smooth specimen appears to be a European representative of the Madagascan *Pachydiscus complanatus*. I know of no other form with which it is likely to be confused. *P.* (*Neodesmoceras*) *obsoletiformis* Jones, 1963 (p. 40, pl. 26, figs. 1–4, 8; pls. 27, 28; text-figs. 20 and 22a) is a Maastrichtian species known only from southern Alaska. It is completely smooth at the diameter of the present specimen.

Occurrence. Assise P^2 in northern Aquitaine; Lower Campanian of Madagascar. An allied form occurs in the Campanian of Northern Ireland (Ulster Museum no. K11558, from Squires Hill, Belfast, County Antrim).

Pachydiscus (*Pachydiscus*) *oldhami* (Sharpe, 1855)

Plate 3; Plate 4, figs. 4 and 5; Plate 5, figs. 1–3; text-figs. 4A, 15, 16, 18

1855 *Ammonites Oldhami* Sharpe, p. 32, pl. 14, fig. 2.
non 1885 *Ammonites Oldhami* Sharpe; Moberg, p. 23, pl. 3, fig. 1.
1894 *Pachydiscus oldhami* Sharpe sp.; de Grossouvre, p. 182, pl. 22, figs. 1 and 6.
1901 *Pachydiscus oldhami* Sharpe; de Grossouvre, p. 281.
1925 *Parapachydiscus Oldhami* Sharpe; Diener, p. 117.
1951 *Pachydiscus oldhami* (Sharpe); Wright and Wright, p. 20.
1974 *Pachydiscus oldhami* (Sharpe, 1855); Naidin, p. 183, pl. 66, figs. 1 and 2; pl. 67, fig. 1; pl. 68, fig. 1.
1980 *Pachydiscus* cf. *oldhami* (Sharpe, 1855); Błaszkiewicz, p. 42, pl. 26, figs. 3 and 4; pl. 37, fig. 4; pl. 49, fig. 2.
non 1982 *Pachydiscus oldhami* (Sharpe, 1854); Tzankov, p. 38, pl. 17, fig. 2.
1984 *Pachydiscus oldhami* (Sharpe, 1855); Kennedy and Summesberger, p. 160.
1985 *Pachydiscus oldhami* (Sharpe); Philip and Bilotte, unnumbered plate.

Type. Holotype by monotypy is the original of Sharpe 1855, p. 32, pl. 14, fig. 2 from the Upper Campanian White Limestone of Dungivan, Northern Ireland, GSM 37244, illustrated here as text-fig. 16.

Material. SP unregistered *ex* Arnaud Collection; 1 specimen labelled Assise R^1, Royan (Le Fort) (Charente-Maritime), but surely from Assise Q; 1 labelled Assise R^1, Gilet-Royan (Charente-Maritime); 1 doubtful specimen from Assise Q, Issac (Dordogne); 1 doubtful specimen from Assise Q, Aubeterre (Charente); 1 doubtful specimen from Assise Q, Neuvic (Gare) (Dordogne). A fine specimen in the EMP collections (Pl. 3) is from Assise Q on the Gironde Estuary; MNHP unregistered, a specimen from Assise Q at Meschers (l'Arnèche) (Gironde); J. P. Platel Collection, a specimen from the top of assise P^3 at Ribérac (Dordogne) (text-fig. 18); Musée de Bordeaux: several specimens from Assise Q at St Georges-de-Didonne (Charente-Maritime); F. Amédro Collection (Calais), a specimen from Bed M of Arnaud (1876) at l'Arnèche, Meschers-sur-Gironde (Charente-Maritime); BMNH C7388 from Assise Q, St Georges de-Didonne (Charente-Maritime).

Dimensions.

	D	*Wb*	*Wh*	*Wh*:*Wb*	*U*
Lectotype	99·0(100)	— (—)	40·0(40·4)	—	32·5(32·8)
at	81·0(100)	— (—)	29·5(36·4)	—	27·0(33·3)
SP unregistered R^1, Royan	55·0(100)	21·0(38·2)	20·5(37·3)	1·02	18·2(33·1)
EMP unregistered	155·0(100)	— (—)	64·0(41·3)	—	49·5(31·9)
F. Amedro Coll.	209·0(100)	— (—)	86·0(41·1)	—	58·0(27·8)
BMNH C7388	208·0(100)	— (—)	83·0(39·9)	—	65·5(31·5)

EXPLANATION OF PLATE 2

Figs. 1 and 2. *Pachydiscus* (*Pachydiscus*) *launayi* de Grossouvre, 1894. SP unregistered *ex* Arnaud Collection, the holotype, from Assise P^1, Voulgézac (Charente).
Figs. 3 and 4. *Hoplitoplacenticeras* (*Hoplitoplacenticeras*) *marroti* (Coquand, 1859). SP unregistered *ex* Arnaud Collection, from Assise P^3 Tauillard (Charente).
Figs. 5 and 6. *Eupachydiscus isculensis* (Redtenbacher, 1873). SP unregistered *ex* Arnaud Collection, from Assise P^1, Montmoreau (Charente).
All figs. × 1.

KENNEDY, *Pachydiscus* (*Pachydiscus*), *Hoplitoplacenticeras* (*Hoplitoplacenticeras*), *Eupachydiscus*

TEXT-FIG. 15. *Pachydiscus* (*Pachydiscus*) *oldhami* (Sharpe, 1855). BMNH C7388, Assise Q, St Georges-de-Didonne (Charente-Maritime), ×0·87.

Description. All specimens are composite moulds, mostly of adults. Early stages up to 55 mm (Pl. 5, figs. 1–3): coiling evolute, with moderately wide umbilicus (32% of diameter), shallow, with a low, rounded wall. Whorl section slightly depressed, with narrowly rounded umbilical shoulder, broadly rounded inner, flattened outer flanks, and broadly rounded venter, the greatest breadth just outside the umbilical shoulder. Twelve to thirteen delicate, narrow primary ribs arise at the umbilical seam. They strengthen into feeble umbilical bullae, or not. These give rise to narrow, distant, rounded rectiradiate ribs that extend to the ventrolateral shoulder before flexing forwards and crossing the venter in a broad convexity. They weaken slightly over the siphonal region and are interrupted at a narrow siphonal groove. One, rarely two secondary ribs separate these primaries; they arise

EXPLANATION OF PLATE 3

Figs. 1–3. *Pachydiscus* (*Pachydiscus*) *oldhami* (Sharpe, 1855). EMP unregistered, from Assise Q on the Gironde Estuary, ×0·95.

KENNEDY, *Pachydiscus* (*Pachydiscus*)

TEXT-FIG. 16. *Pachydiscus* (*Pachydiscus*) *oldhami* (Sharpe, 1855). The holotype, GSM 37244, the original of Sharpe 1855, pl. 14, fig. 2, Upper Campanian White Limestone of Dungivan, northern Ireland, × 1.

at or below mid-flank, are initially weaker than, but eventually match the primaries, which they parallel, giving a total of twenty-two to twenty-four ribs per half whorl.

One specimen shows the middle growth stages (Pl. 4, figs. 4 and 5), where similar differentiation into primary and secondary ribs persists, although the inner flank ornament effaces progressively. The outer whorl of adult phragmocones (Pl. 3; text-figs. 15 and 18) shows a progressive decline of ornament, and, ultimately, ornament cannot be distinguished from the irregular surface of the mould (text-fig. 18).

Sutures complex, typical for genus.

Discussion. The holotype of *Pachydiscus* (*Pachydiscus*) *oldhami* is shown in text-fig. 16. The surface is badly corroded, and it bears only passing resemblance to Sharpe's figure (1855, pl. 14, fig. 2). This species is a close ally of *P*. (*P*.) *haldemsis* (Schlüter, 1867), based on a microconch (lectotype refigured here as text-fig. 11A, B), of which *P*. (*P*.) *koeneni* de Grossouvre, 1894 (p. 178; see revision in Kennedy and Summesberger 1984) is the macroconch and a synonym. Microconchs of *oldhami* have yet to be confidently recognized. The macroconchs differ in the generally coarser ornament of *haldemsis* (compare Pl. 5, figs. 1–3 and 7–14) with well-developed rather than incipient bullae when young. Adult phragmocones differ in the persistence of ornament in *haldemsis* whereas it effaces in *oldhami*, notably on the inner flank (compare text-figs. 15 and 17). Adult *oldhami* are smooth (Naidin 1974, pl. 46, pl. 67, fig. 1) whereas *haldemsis* retain strong ventral and inner flank ornament (Błaskiewicz 1980, pl. 27; Kennedy and Summesberger 1984, pl. 6, fig. 2; pl. 14, fig. 1). In Poland and Aquitaine *haldemsis* precedes and overlaps with *oldhami*, and they may be no more than extremes of one variable species. I also suspect but cannot prove that the paralectotype of *P*. *ambiguus* de Grossouvre, 1894 (p. 198, pl. 29, fig. 3) from Assise P^3 at Tauillard (Charente) is a microconch of *oldhami*; it is described and discussed below as *P.* (*Pachydiscus*) sp.

Occurrence. Upper Campanian. Top of Assise P^3 and Q in northern Aquitaine; Tercis (Landes); Northern Ireland; Norfolk, England; Poland; European Russia.

Pachydiscus (*Pachydiscus*) *haldemsis* (Schlüter, 1867)

Plate 4, figs. 1–3; Plate 5, figs. 7–14; text-figs. 11A–D, F, G and 17

1867 *Ammonites haldemsis* Schlüter, p. 19, pl. 3, fig. 1.
1872a *Ammonites haldemsis* Schlüter, p. 70.
1872a *Ammonites* cf. *auritocostatus* Schlüter, p. 70, pl. 22, figs. 6 and 7.
1872a *Ammonites Galicianus* Favre; Schlüter, p. 63, pl. 19, figs. 3–5.
1885 *Ammonites Oldhami* Sharpe; Moberg, p. 23, pl. 3, fig. 1.
1889 *Pachydiscus galicianus* Favre; Griepenkerl, p. 101.
1894 *Pachydiscus koeneni* de Grossouvre, p. 178.
1902 *Pachydiscus galicianus* (Schlüter); Wollemann, p. 103.
1913 *Pachydiscus oldhami* Sharpe; Nowak, p. 362, pl. 41, fig. 16; pl. 43, fig. 31; pl. 45, fig. 43.
?1913 *Pachydiscus kaliszanensis* Nowak, p. 359, pl. 40, fig. 8.
1913 *Pachydiscus haldemensis* Schlüter; Nowak, p. 349.
1925 *Pachydiscus koeneni* Grossouvre; Diener, p. 106.
1935 *Pachydiscus* ex. aff. *galicianum* Favre; Brinkmann, p. 5.
1951 *Pachydiscus* cf. *Koeneni* Grossouvre; Mikhailov, p. 60, pl. 10, fig. 47.
1951 *Pachydiscus oldhami* (Sharpe); Wright and Wright, p. 20 (*pars*)
1955b *Pachydiscus haldemensis* Schlüter; Matsumoto, p. 168.
1955b *Pachydiscus ambiguus* Grossouvre; Matsumoto, p. 169.
1957 *Pseudomenuites ambiguus* (de Grossouvre); Wright, p. 380.
1959 *Pachydiscus koeneni* Grossouvre; Naidin and Shimanskij, p. 185, pl. 9, fig. 1.
1964 *Pachydiscus koeneni* Grossouvre; Giers, p. 263, pl. 5, figs. 1 and 2; text-fig. 5.
1974 *Menuites ambiguus* (Grossouvre, 1894); Naidin, p. 182, pl. 63, fig. 2.
1974 *Pachydiscus koeneni* Grossouvre, 1894; Naidin, p. 186, pl. 65, figs. 2 and 3.
1980 *Pachydiscus koeneni* Grossouvre, 1894; Błaszkiewicz, p. 42, pl. 26, figs. 1 and 2; pl. 27, figs. 1–4; pl. 28, figs. 1–4; pl. 34, figs. 3 and 4.
1984 *Pachydiscus* (*Pachydiscus*) *haldemsis* (Schlüter, 1867); Kennedy and Summesberger, p. 158, pl. 4, figs. 1–5; pl. 5, fig. 1; pl. 6, fig. 2; pl. 7, figs. 1–11; pl. 13, fig. 1 (with additional synonymy).

Types. The lectotype of *Ammonites haldemsis* Schlüter, 1867 (p. 19) by the subsequent designation of Kennedy and Summesberger 1984, p. 158, is an unregistered specimen in the Schlüter Collection (GPIB), the original of Schlüter 1867, pl. 3, fig. 1, refigured here as text-fig. 11A–B. The lectotype of *P. koeneni* de Grossouvre, 1894 (p. 178) is in the same collection and is shown in text-fig. 17. Both are from the Upper Campanian of Haldem, Westphalia.

Material. SP unregistered *ex* Arnaud Collection; five specimens from Assise P^3, St Médard-de-Barbezieux (Charente). One from the same horizon at Celles (Dordogne) in the Sinclair Collection (Allemans).

Discussion. The suite of specimens from St Médard-de-Barbezieux closely resembles juvenile macroconchs described in detail by Kennedy and Summesberger (1984) and match such of the innermost whorls as are visible within the lectotype (text-fig. 17). The species closely resembles *P.* (*P.*) *oldhami* (Sharpe, 1855); differences between the two are noted above. The Celles specimen shown in Plate 4, figs. 1–3 has rather weaker ribbing than the St Médard specimens, combining some features of the ornament of *P.* (*P.*) *oldhami* (text-fig. 16) with the coiling typical of *P.* (*P.*) *koeneni*.

The large specimen from Assise Q at Saint-Georges-de-Didonne figured by de Grossouvre (1894, pl. 22, fig. 1; as *oldhami*) was referred to *koeneni* (= *haldemsis*) by Błaszkiewicz (1980) and may be better referred there than to *oldhami*. I have not seen the original, however, and other specimens from the locality (text-fig. 15) seem better referred to Sharpe's species. As noted previously (Błaszkiewicz 1980; Kennedy and Summesberger 1984) the two are closely related.

Occurrence. Upper Campanian. Assise P^3 and possible Q in Aquitaine; north Germany, Poland, Sweden, Donbass, Kopet Dag, and elsewhere in the USSR. Possibly present in the White Limestone in Northern Ireland.

TEXT-FIG. 17. *Pachydiscus* (*Pachydiscus*) *haldemsis* (Schlüter, 1867). The holotype of *P.* (*P.*) *koeneni* de Grossouvre, 1894, GPIB unregistered, Upper Campanian of Haldem, Westphalia; the macroconch of *P.* (*P.*) *haldemsis*, × 1.

EXPLANATION OF PLATE 4

Figs. 1–3. *Pachydiscus* (*Pachydiscus*) *haldemsis* (Schlüter, 1867). Sinclair Collection, from Assise P^3, Celles (Dordogne), × 1·5.

Figs. 4 and 5. *P.* (*P.*) *oldhami* (Sharpe, 1855). MNHP unregistered, from Assise Q, Meschers (l'Arnèche) (Gironde), × 0·8.

1

2

3

4

5

KENNEDY, *Pachydiscus* (*Pachydiscus*)

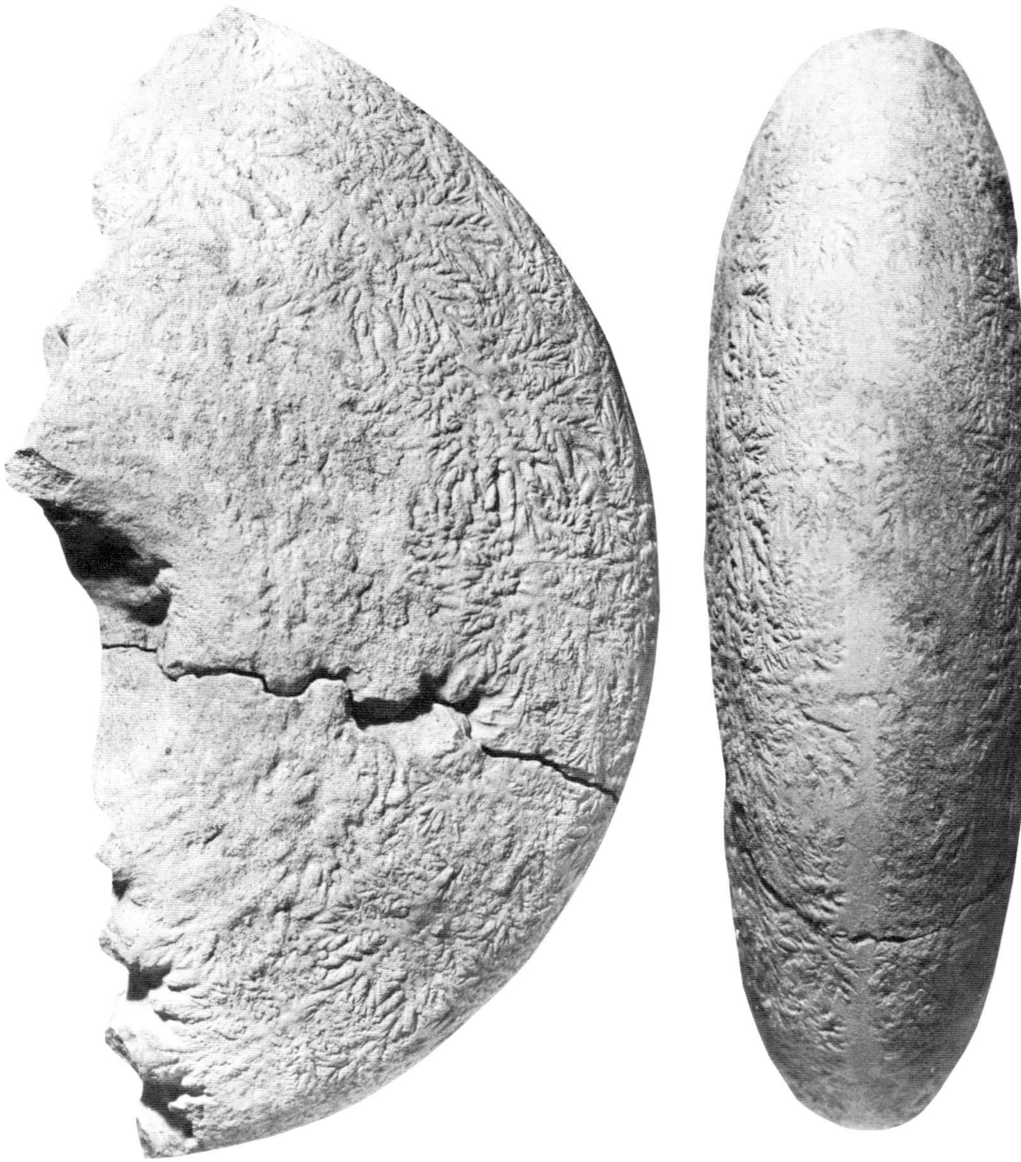

TEXT-FIG. 18. *Pachydiscus* (*Pachydiscus*) *oldhami* (Sharpe, 1855), Upper Campanian, top of Assise P³ of Arnaud at Ribérac (Dordogne); J. P. Platel Collection, ×0·7.

EXPLANATION OF PLATE 5

Figs. 1–3. *Pachydiscus* (*Pachydiscus*) *oldhami* (Sharpe, 1855). SP unregistered *ex* Arnaud Collection, from Assise R¹, Royan (Le Fort) (Charente-Maritime).

Figs. 4–6. *P.* (*Pachydiscus*) sp. The paralectotype of *P. ambiguus* de Grossouvre, 1894, pl. 29, fig. 3, SP unregistered *ex* Arnaud Collection, from Tauillard (Charente).

Figs. 7–14. *P.* (*P.*) *haldemsis* (Schlüter, 1867). SP unregistered *ex* Arnaud Collection, from Assise P³, St Médard-de-Barbezieux (Charente).

All figs. ×1.

PLATE 5

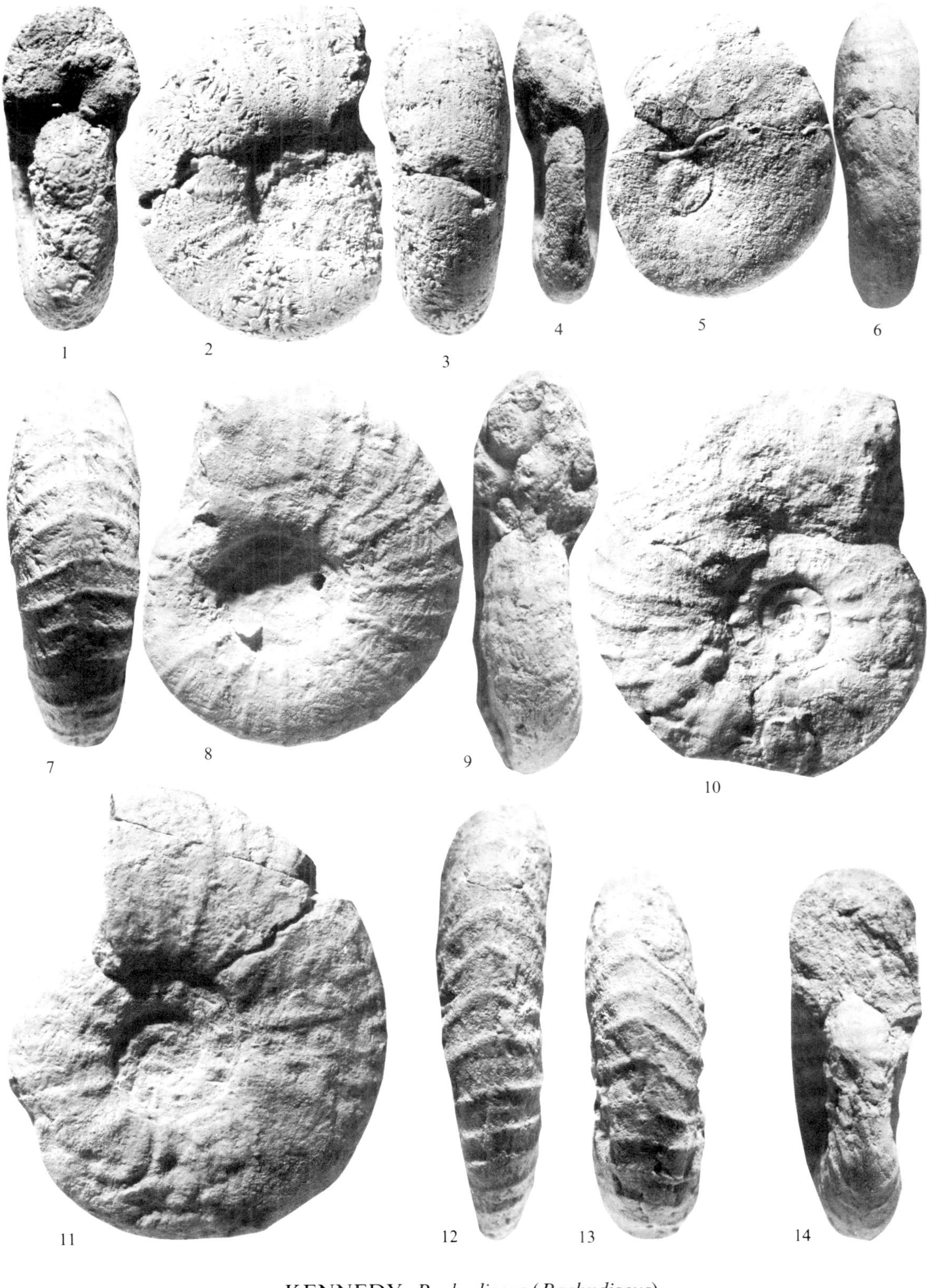

KENNEDY, *Pachydiscus* (*Pachydiscus*)

Pachydiscus (*Pachydiscus*) sp.

Plate 5, figs. 4–6; Plate 6, fig. 5

1894 *Pachydiscus ambiguus* de Grossouvre, p. 198 (*pars*), pl. 29, fig. 3 only.
1922*a* '*Pachydiscus*' *ambiguus* de Grossouvre; Spath, p. 123 (*pars*).
1925 *Menuites ambiguus* Grossouvre; Diener, p. 112 (*pars*).
1955*b* *Pachydiscus ambiguus* Grossouvre; Matsumoto, p. 169.
1957 *Pseudomenuites ambiguus* (de Grossouvre); Wright, p. 380 (*pars*).
1974 *Menuites ambiguus* (Grossouvre, 1894); Naidin, p. 182, pl. 63, fig. 2.
1984 *Pachydiscus ambiguus* de Grossouvre; Kennedy and Summesberger, p. 160.

Material. SP unregistered *ex* Arnaud Collection, the original of de Grossouvre 1894, pl. 29, fig. 3, the paralectotype of *Pachydiscus ambiguus* de Grossouvre, 1894, from Assise P^3, Tauillard (Charente).

Description. The only available specimen is a small crushed composite mould. Coiling involute, with 70% of previous whorl covered. Umbilicus small (25% of diameter), shallow, with flattened wall. Because of crushing original whorl section cannot now be determined. Ornament is visible on the last half whorl only; the early parts are badly worn. There are an estimated ten primary ribs per whorl. They are weak on the umbilical wall but strengthen at the shoulder. They are narrow, rounded, and very distant, straight and prorsiradiate on the flank, projected forwards over the ventrolateral shoulder, crossing the venter in a broad convexity. Occasional secondaries intercalated around mid-flank are also developed. On the last third of the outer whorl the ribs develop prominent ventral tubercles.

Sutures not seen.

Discussion. De Grossouvre based his *P. ambiguus* on this specimen and the figure of *A. cf. auritocostatus* Schlüter, 1872*a* (p. 70, pl. 22, figs. 6 and 7). *A. auritocostatus* is an *Anapachydiscus* microconch, as noted above, while Diener (1925, p. 112) designated Schlüter's specimen lectotype of *ambiguus*. (Matsumoto (1955*b*, p. 169, footnote) designated the present specimen lectotype of *ambiguus*, overlooking Diener's prior action.) As can be seen from comparing this specimen (text-fig. 11F and G) with the lectotype and other specimens of microconch *P.* (*P.*) *haldemsis* (text-fig. 11A–D), the lectotype of *ambiguus* is simply a crushed *haldemsis* body-chamber. The present specimen represents a different form with more primary ribs, ten versus seven to eight per half whorl at the same ontogenetic stage. I suspect it is a microconch *P.* (*P.*) *oldhami*, but lack good evidence at this time.

Occurrence. As for material.

Genus EUPACHYDISCUS Spath, 1922*a*, p. 124
[= *Mesopachydiscus* Yabe and Shimizu, 1926, p. 172]

Type species. Ammonites isculensis Redtenbacher, 1873, p. 122, pl. 29, fig. 1, by original designation.

Discussion. The type species of *Eupachydiscus* is based on a macroconch. Immel *et al.* (1982) have recently described, as *Pseudomenuites katschthaleri* (p. 20, pl. 7, figs. 2–5), a small bituberculate form

---

EXPLANATION OF PLATE 6

Fig. 1. *Nostoceras* (*Bostrychoceras*) *polyplocum* (Roemer, 1841). P. Moreau Collection (Poitiers), from Assise P^3, Zone Industrielle, Sain-Georges, Royan (Charente-Maritime).
Fig. 2. *Pseudophyllites indra* (Forbes, 1846), SP unregistered *ex* Arnaud Collection, from Assise P^3, Petingaud (Charente).
Figs. 3 and 4. *Submortoniceras* sp. SP unlocalized, by its preservation from Assise P of northern Aquitaine.
Fig. 5. *Pachydiscus* (*Pachydiscus*) sp. The paralectotype of *P. ambiguus* de Grossouvre, 1894, pl. 29, fig. 3, SP unregistered *ex* Arnaud Collection, from Tauillard (Charente).
Figs. 6–8. *P.* (*P.*) cf. *complanatus* Collignon, 1952. SP unregistered *ex* Arnaud Collection, from Assise P^2, St-Laurent-de-Belzagot (Charente).
All figs. ×1.

KENNEDY, *Nostoceras* (*Bostrychoceras*), *Pseudophyllites*, *Submortoniceras*, *Pachydiscus* (*Pachydiscus*)

that may be the microconch of the type species. *Pseudomenuites* Matsumoto, 1955*b* (p. 169) has *Pachydiscus ambiguus* de Grossouvre, 1894 as type species; this is a microconch *P.* (*Pachydiscus*) (see above).

Eupachydiscus isculensis (Redtenbacher, 1873)

Plate 2, figs. 5 and 6; Plate 10, figs. 13 and 14; text-fig. 19

1873 *Ammonites isculensis* Redtenbacher, p. 122, pl. 29, fig. 1.
1894 *Pachydiscus isculensis* Redtenbacher sp.; de Grossouvre, p. 185, pl. 22, fig. 2; pl. 26, fig. 1; pl. 37, fig. 1.
?1898 *Pachydiscus isculensis* Redtenbacher sp.; Mariani, p. 55; text-fig. on p. 56.
1906 *Pachydiscus isculensis* Redtenbacher; Müller and Wollemann, p. 9, pl. 7, figs. 1 and 2; text-figs. 1 and 2.
1922*a* *Eupachydiscus isculensis* (Redtenbacher); Spath, p. 124.
1925 *Eupachydiscus isculensis* Redtenbacher; Diener, p. 110.
1931 *Pachydiscus isculensis* (Redtenbacher); Basse, p. 34.
1932 *Pachydiscus isculensis* (Redtenbacher); Collignon, p. 51, pl. 3, fig. 4; text-fig. 5.
1935 *Pachydiscus isculensis* (Redtenbacher); Brinkmann, p. 4.
?1938 *Eupachydiscus isculensis* (Redtenbacher); Dietrich, pp. 228 ff.
?1938 *Eupachydiscus isculensis* (Redtenbacher); Collignon, p. 27.
1950 *Eupachydiscus isculensis* (Redtenbacher); Hourcq, pp. 72 and 78.
1952 *Eupachydiscus isculensis* Redt.; Collignon, p. 31, pl. 4, fig. 1; pl. 6, fig. 1.
1955 *Eupachydiscus isculensis* Redt.; Collignon, p. 32, pl. 4, fig. 1; pl. 6, fig. 1; text-fig. 4.
1957 *Eupachydiscus isculensis* (Redtenbacher); Wright, p. 380, fig. 495, 2.
1961 *Eupachydiscus isculensis* (Redtenbacher); Gerth, p. 124, pl. 24, fig. 7.
1966 *Eupachydiscus isculensis* (Redtenbacher); Collignon, ?p. 36, pl. 470, fig. 1918; ?pl. 471, fig. 1918; p. 90, pl. 492, fig. 1977.
1979 *Eupachydiscus isculensis* (Redtenbacher); Wiedmann *in* Herm *et al.*, p. 49, pl. 8A.
1979 *Eupachydiscus isculensis* (Redtenbacher); Summesberger, p. 136, pl. 7, figs. 30 and 31; pl. 8, fig. 32; text-figs. 20, 21, 22.
1982 *Eupachydiscus isculensis* (Redtenbacher); Martinez, p. 83, pl. 7, fig. 3.
1982 *Eupachydiscus isculensis* (Redtenbacher, 1873); Tzankov, p. 39, pl. 17, fig. 3; pl. 18, figs. 2 and 3.
?1982 *Pseudomenuites katschthaleri* Immel *et al.*, p. 20, pl. 7, figs. 2–5 (with synonymy).
1982 *Eupachydiscus isculensis* (Redtenbacher, 1873); Immel *et al.*, p. 22, pl. 7, figs. 8–11; pl. 8, figs. 1–4.
1983 *Eupachydiscus isculensis* (Redt.); Collignon *in* Bilotte and Collignon, p. 194.

Holotype. By monotypy, an unregistered specimen in the Oberösterreichisches Landesmuseum, Austria, from the Santonian Gosau Beds between Ischl and Kohlbuch, Austria, the original of Redtenbacher 1873, pl. 29, fig. 1, reillustrated here as text-fig. 19.

Material. SP unregistered *ex* Arnaud Collection, two specimens: one from Assise P[1], Éraville (Charente), the original of de Grossouvre 1894, pl. 22, fig. 2; one from the same horizon at Montmoreau (Charente).

Description. The largest specimen is a juvenile 47 mm in diameter. Coiling is moderately involute, with a deep umbilicus. The whorl section is depressed reniform, with the greatest breadth at the umbilical bullae. There are sixteen ribs on the last half whorl. They arise singly or in pairs from pronounced umbilical bullae, while non-bullate primaries and intercalatory ribs are inserted between the bullate ones. The ribs are straight and prorsiradiate on the inner flank but project forwards over the ventrolateral shoulder to cross the venter in a broad convexity. They are all very narrow and sharp, and are separated by wide interspaces.

Sutures not seen.

Discussion. The much larger holotype (text-fig. 19) is a magnificent adult specimen with just over half a whorl of body-chamber, retaining extensive traces of the original aragonitic shell. It is somewhat crushed, and deformed into an ellipse with major dimension of 190 mm. Coiling is moderately involute, with approximately 70% of the previous whorl being covered. The umbilicus is deep and narrow (approximately 20–25% of the diameter), with a high, outwards-inclined wall. The whorl section

TEXT-FIG. 19. *Eupachydiscus isculensis* (Redtenbacher, 1873). The holotype, an unregistered specimen in the Oberösterreichisches Landesmuseum, Austria, from the Santonian Gosau Beds between Ischl and Kohlbuch, Austria, the original of Redtenbacher 1873, pl. 29, fig. 1, ×0·75.

appears to have originally been depressed, with an abruptly rounded umbilical shoulder, rounded convergent flanks, and rounded venter, the greatest breadth being at the umbilical shoulder.

At the smallest diameters visible, about 12 strong ribs arise on the umbilical wall, and develop into strong bullae at the umbilical shoulder, which in turn give rise to coarse, prorsiradiate primary ribs. On the outer whorl, there are 17 primary ribs which arise at the umbilical seam and pass back across the umbilical wall, strengthening as they do so, and developing into massive umbilical bullae on the umbilical shoulder. At the beginning of the outer whorl, strong ribs arise in groups of 2 or 3 from these bullae, but as size increases, the ribs are single, strong, and prorsiradiate, straight on the inner flank but concave across the middle and outer flank, whence they sweep across the venter. Secondary ribs intercalate at various positions on the flank, with generally 1 or 2 between the pairs of primary ribs, so that there is a total of 42–43 ribs on the outer whorl. At the aperture the ribs decline markedly after a final strong bullate primary. This specimen can be linked to the Aquitaine specimen, via a series of well-preserved specimens recently described from the Gosau Group by Summesberger (1979) and Immel *et al.* (1982). The latter figure, as *Pseudomenuites katschthaleri* Immel, Klinger and Wiedmann, 1982 (p. 20, pl. 7, figs. 2–5) a small form which has early whorls like those of *isculensis* (compare pl. 7, fig. 2 and pl. 7, fig. 8 in Immel *et al.*) but with ventral tubercles on the body-chamber linking pairs of flank ribs which thus loop between these and the umbilical bullae. As noted elsewhere (p. 34), bituberculate pachydiscids are in some cases demonstrably microconchs, and I take it to be the case with *katschthaleri* vs. *isculensis*.

E. grossouvrei (Kossmat, 1898) (p. 101 (166)) = *Ammonites Ootacodensis* Stoliczka, 1865, p. 109, pl. 57, is a mid-Campanian species, easily distinguished by its much more delicate ornament with more numerous intercalatories, as in *E. pseudogrossouvrei* Collignon, 1952 (p. 42, pl. 8, figs. 1 and 2; 1955, p. 42, pl. 8, figs. 1 and 2) from the mid-Campanian of Madagascar. *E. haradai* (Jimbo, 1894) (p. 29, pl. 2, fig. 2; see Usher 1952, p. 73, pl. 12, figs. 2–4; pl. 13, figs. 1–3; pl. 31, fig. 6) is densely ribbed with weak bullae. *E. jeani* (de Grossouvre, 1894) (p. 187, pl. 26, fig. 5), from the Santonian of Aude, is a curious species: the inner whorl bears fine, dense, crowded straight ribs without bullae, and the ribs coarsen abruptly on the later parts of the outer whorl, but without the prominent bullae of *E. isculensis*. In *Pachydiscus launayi* de Grossouvre, 1894, a Campanian species described above and referred to *Eupachydiscus* by some authors, ribbing is finer, denser, without bullae (Pl. 2, figs. 1 and 2). *E. sayni* (de Grossouvre, 1894) (p. 181, pl. 29, fig. 2) may be similarly distinguished. Other species, listed by Collignon (1952, p. 86; 1955, p. 79) may also be easily separated from the present form.

Occurrence. E. isculensis ranges from mid-Santonian to Lower Campanian. It is known from Assise P^1 in northern Aquitaine, the Santonian of the Corbières, northern Spain, Austria, Roumania, and Zululand, and the Santonian and Campanian of Madagascar.

Genus BAYLEITES Collignon, 1952, p. 22

Type species. Bayleites baylei Collignon, 1952, p. 22, pl. 9, fig. 2.

Discussion. Collignon based the type species on a single specimen in the École des Mines Collections, said to be from Talmont (Charente-Maritime). Re-examination of the specimen shows that its preservation is not that of the Upper Cretaceous of northern Aquitaine but rather that of the Middle Bathonian of Lucon (Vendée). The specimen has been examined by Professor J. H. Callomon (London) and colleagues, who considered it a *Cadomites orbignyi* de Grossouvre, 1930 (p. 373, pl. 39, fig. 6; see also Arkell 1952, p. 80).

Family MUNIERICERATIDAE Wright, 1952, p. 222
[= Pseudoschloenbachiidae Van Hoepen, 1968, p. 186]

Genus PRAEMUNIERICERAS Collignon, 1966, p. 13

Type species. Praemuniericeras proprium Collignon, 1966, p. 14, pl. 460, fig. 1882.

EXPLANATION OF PLATE 7

Figs. 1–5. *Hoplitoplacenticeras* (*Hoplitoplacenticeras*) *rejaudryi* (de Grossouvre, 1894). 1–3, holotype; 4–5, holotype of *Sonneratia rara* de Grossouvre, 1894, SP unregistered *ex* Arnaud Collection, from Assise P^3, St-Médard-de-Barbezieux (Charente).

Figs. 6 and 7. *Pachydiscus* (*Pachydiscus*) *launayi* de Grossouvre, 1894, SP unregistered *ex* Arnaud Collection, from Assise P^1, Montmoreau (Charente).

All figs. ×1.

PLATE 7

KENNEDY, *Hoplitoplacenticeras* (*Hoplitoplacenticeras*), *Packydiscus* (*Pachydiscus*)

Praemuniericeras? sp.

Plate 11, figs. 5 and 6

Material. MNHP B17226, from Assise P of Arnaud, labelled La Vaure, near Montignac (Dordogne).

Description. The specimen is a fragment of body-chamber only. Coiling appears to have been relatively evolute, with a compressed whorl section, the flanks flattened and convergent, the ventrolateral shoulders broadly rounded and the venter bluntly fastigiate. Ornament consists of narrow rounded straight primaries which sweep forwards across the ventrolateral shoulders and terminate in broad blunt spatulate endings that form an obtuse chevron on the venter. The apex of the chevrons is bisected by a siphonal ridge bearing elongate siphonal clavi.

Discussion. This specimen closely resembles the diverse *Praemuniericeras* described by Collignon (1966) from the Santonian of Madagascar, and is tentatively referred to this genus, not previously recorded from Europe or from the Campanian.

Superfamily HOPLITACEAE H. Douvillé, 1890, p. 290

[*nom. correct.* Wright and Wright 1951, p. 21 (*pro* Hoplitida Spath, 1922*b*, p. 95, *nom. transl. ex.* Hoplitida Douvillé, 1890) (= Placenticeratacea Hyatt, 1900, p. 584, *nom. correct.* Casey, 1960, p. 208, *pro* Placenticeratida Hyatt, 1900; Engonocerataceae Hyatt, 1900, p. 585, *nom. transl.* Basse, 1952, p. 658, *ex* Engoniceratidae Hyatt, 1900)]

Family PLACENTICERATIDAE Hyatt, 1900, p. 585

[= Hypengonoceratinae Chiplonkar and Ghare, 1976, p. 2; Baghiceratinae Chiplonkar and Ghare, 1976, p. 3]

Genus PLACENTICERAS Meek, 1876*a*, p. 462

[for synonymy see Kennedy and Wright 1983, p. 869]

Type species. Ammonites placenta DeKay, 1828, p. 278, by original designation by Meek 1876*a*, p. 426.

Discussion. See Kennedy and Wright (1983) for diagnosis, discussion of *Placenticeras*, and reasons for regarding the mainstream of the family as a single genus. Those authors separated *Karamaites* Sokolov, *in* Casey 1965 (p. 461) (type species *Placenticeras grossouvrei* Semenov, 1899, p. 97) largely because of the absence of clear dimorphism. Subsequently Seyed-Emami *et al.* (1984, p. 163, figs. 3, 1–2; 4; 5, 1, 2; 6, 1, 2), illustrated fragments of both weakly and strongly ornamented specimens up to 360 mm diameter (and still septate) that cast further doubt on separation.

Study of specimens of *P. bidorsatum* from Broitzem (text-fig. 20) in the Bonn Collections shows that in some species at least the ventral clavi are in fact the bases of long septate spines, a feature not previously recognized in the family.

Placenticeras bidorsatum (Roemer, 1841)

Plate 8, figs. 1–12; Plate 14, figs. 8 and 9; text-figs. 20–23

1841 *Ammonites bidorsatus* Roemer, p. 88, pl. 13, fig. 5.
1867 *Ammonites polyopsis* Dujardin; Schlüter, p. 25, pl. 4, figs. 1 and 2.
1872*a* *Ammonites bidorsatus* Röm.; Schlüter, p. 51, pl. 15, figs. 6–8.
1894 *Placenticeras bidorsatum* A. Römer sp.; de Grossouvre, p. 137, text-fig. 55.
1900 *Diplacomoceras bidorsatum* Schlüter; Hyatt, p. 585.
1903 *Diplacmoceras bidorsatum* (Roemer); Hyatt, p. 243.

EXPLANATION OF PLATE 8

Figs. 1–12. *Placenticeras bidorsatum* (Roemer, 1841). All SP unregistered *ex* Arnaud Collection. 1–7 are from Assise P[1], Livernant (Charente); 8–10 from Assize P[1], Mensignac-Vaure (Dordogne); 11 and 12 are labelled P[2], Montmoreau (Charente).
All figs. × 1.

KENNEDY, *Placenticeras*

1903 *Diplacmoceras canaliculatum* n. sp. Hyatt, p. 243.
?1905 *Placenticeras bidorsatum* A. Roemer; Wegner, p. 209.
1906 *Placenticeras bidorsatum* A. Roemer; Müller and Wolleman, p. 6, pl. 3, figs. 1 and 2; *non* pl. 4, fig. 5; pl. 9, figs. 1 and 2.
1906 *Placenticeras bidorsatum* A. Roemer var. *glaberrima* Müller and Wollemann, p. 7, pl. 3, fig. 3; pl. 9, fig. 2.
1925 *Diplacmoceras bidorsatum* Roemer; Diener, p. 191.
1925 *Diplacmoceras canaliculatum* Hyatt; Diener, p. 191.
1925 *Diplacmoceras bidorsatum* var. *globerrima* Diener, p. 191.
1931 *Diplacmoceras bidorsatum* A. Röm.; Riedel, p. 697.
1931 *Diplacmoceras bidorsatum* A. Röm. var. *glaberrima* G. Mull. and A. Wollem.; Riedel, p. 698.
1957 *Diplacmoceras bidorsatum* (Roemer); Wright, p. 392, fig. 483, 2.
1985 *Placenticeras* (*Diplacomoceras*) *bidorsatum* (Schlüter, 1867); Kennedy, pl. 3, figs. 1 and 2.

Types. Roemer cites specimens from Dülmen and Blankenburg. None survive in the Roemer Collection in Wrocław, and in the interests of nomenclatural stability I here designate as neotype GPIB 46, the original of Schlüter 1867, pl. 4, fig. 1 and 1872*a*, pl. 15, figs. 7 and 8 (text-fig. 21), from the Lower Campanian of Dülmen, Westphalia, German Federal Republic, one of the localities of the now missing syntypes. This is the lectotype, here designated, of *Diplacmoceras canaliculatum* Hyatt, 1903 (p. 243) which becomes an objective synonym of *bidorsatum*. The paralectotype of *canaliculatum* is GPIB 24, the original of Schlüter 1867, pl. 4, fig. 2 and 1872*a*, pl. 15, fig. 6 (text-fig. 22).

Material. SP unregistered *ex* Arnaud Collection: 6 specimens from Assise P^1, Livernant (Charente); 1 from Assise P^3, Tauillard (Charente); 1 from P^1 and 1 from P^{1-2}, Mensignac-Vaure (Dordogne); 1 from P^2 inf., St Hilaire (Charente); 1 from P^2, La Côte (Charente); 1 from P^2, Montmoreau (Charente).

Dimensions.

	D	*Wb*	*Wh*	*Wb*:*Wh*	*U*
Livernant 1	55·8(100)	— (—)	32·3(57·9)	—	7·8(14·0)
Livernant 2	75·0(100)	17·6(23·5)	38·5(51·3)	0·46	11·0(14·7)
Montmoreau P^{1-2}	73·2(100)	17·5(23·9)	37·0(50·5)	0·47	12·0(16·4)
St Hilaire P^2	82·5(100)	18·5(22·4)	41·5(50·3)	0·45	11·6(14·1)
GPIB 46 at:	120·0(100)	— (—)	59·5(49·6)	—	19·5(16·3)
GPIB 24	140·0(100)	35·5(25·3)	71·2(50·9)	0·50	20·2(14·4)
Roemer, pl. 13, fig. 5	(100)	(28·9)	(50·2)	0·57	(18·2)

Description. All specimens are internal moulds or composite moulds of juveniles ranging from 57–97 mm in diameter. Very involute, up to 80% of previous whorl covered. Umbilicus crater-like, small (14–16% of diameter), shallow, low umbilical wall flattened and sloping outwards; umbilical shoulder narrowly rounded, sharp. Whorl section very compressed (Wb: Wh 0·45–0·5), greatest breadth below mid-flank. Inner flanks broadly rounded, outer flanks flattened and convergent. Venter very narrow, concave with sharp ventral shoulders when well preserved, flat or rounded in some specimens. Moulds are smooth (Pl. 8, figs. 2, 7, 9) in most cases; a few bear faint crescentic concave ribs on outer flank (Pl. 8, fig. 4; Pl. 14, figs. 8 and 9).

Suture poorly exposed; eight to nine saddles in external part (Pl. 8, fig. 9).

Discussion. The Aquitaine specimens are all juveniles, but show whorl proportions, suture and ventral morphology that links them to the much larger neotype (text-fig. 21). They are the smallest specimens yet described that can be attributed to the species and are of interest in that some show crescentic flank ornament so typical of compressed variants of older *Placenticeras* species (e.g. de Grossouvre 1894, pl. 5, fig. 2). The neotype itself is a composite internal mould of a near-complete adult. Coiling is involute, with 67% of the previous whorl covered at the largest diameter preserved, where the umbilical seam is egressing markedly, indicating maturity. The surface of the phragmocone is badly preserved, with the last suture at a diameter of 97·5 mm. No ornament is visible below a diameter of 75–80 mm. At this point, long, low, feeble ventral clavi appear on the ventrolateral shoulder, followed by feeble umbilical bullae and distant, conical outer lateral tubercles. Initially weak, this ornament strengthens progressively around the body-chamber. The bullae give rise to narrow, delicate,

TEXT-FIG. 20. *Placenticeras bidorsatum* (Roemer, 1841). GPIB unregistered *ex* Schlüter Collection, Lower Campanian of Broitzem, Braunschweig, German Federal Republic. Note the remarkable preservation of septate spines, × 1.

prorsiradiate ribs which bear a tiny inner lateral bulla; this migrates progressively outwards as the aperture is approached. The bullae give rise to low, broad ribs, singly or in pairs, and these flex back over the outer flank, where they are convex, and accompanied by irregular striae, linking in groups to the outer lateral tubercles. Weak on the phragmocone, these strengthen and migrate out to an inner ventrolateral position on the body-chamber and change from conical to a bulla that has a steep apical termination but projects obliquely forwards to the ventrolateral shoulder in a rib-like process. The ventrolateral shoulders remain sharp and keel-like on either side of the concave venter and are undulose, bearing long, low clavi, approximately three times more numerous than the inner row of tubercles, although at times scarcely discernible. The outer lateral–inner ventrolateral tubercles on the last half of the body-chamber bear, on their apertural side, groups of ridges and grooves that are concave with respect to the line of the venter and come close to linking successive tubercles in a festoon-like fashion. Schlüter's second specimen is still septate at 140 mm diameter. The surface is very

TEXT-FIG. 21. *Placenticeras bidorsatum* (Roemer, 1841). Neotype, GPIB 24, the original of Schlüter 1867, pl. 4, fig. 1 and 1872*a*, pl. 15, figs. 7 and 8, also the lectotype of *P. canaliculatum* Hyatt, 1903, Lower Campanian of Dülmen, Westphalia, German Federal Republic, × 0·78.

poorly preserved, but shows the same style of tuberculation as the neotype, although umbilical and lateral tubercles are coarser (text-fig. 22).

Variation in adults of this species is documented by specimens from Broitzem, Braunschweig, figured by Müller and Wollemann (1906) as well as additional specimens from this locality in the Schlüter Collection (GPIB unregistered). These include small individuals, adult at 95–110 mm. These have stout, subrectangular whorl sections (e.g. text-fig. 23; Müller and Wollemann 1906, pl. 9, fig. 1) and are microconchs, much smaller than Schlüter's macroconch, still septate at 140 mm. This type of dimorphism is exactly comparable to that seen in the Santonian *P. polyopsis* (Dujardin, 1837) (*fide* Kennedy and Wright 1983) and other species where '*Stantonoceras*' microconchs and *Placenticeras* macroconchs occur consistently together. The Broitzem material also shows a degree of variation in strength of ribbing and tuberculation in the species. Most remarkable of all is the occurrence of material which shows the outer lateral–inner ventrolateral tubercles to be the bases only of long septate spines (text-fig. 20). This same phenomenon is also seen in *Hoplitoplacenticeras* species (see below), but I know of no other occurrence in *Placenticeras*.

Hyatt (1903) separated Schlüter's specimens from Roemer's, renaming the former *canaliculatum*. Most authors have taken Schlüter's illustration as typical of the species. They were differentiated because of the stouter whorl section in Roemer's figure (reproduced here as text-fig. 24), broader venter and the presence of narrow inner-flank folds. These 'inner folds', which are shown bearing tiny inner lateral tubercles are also shown by Schlüter's material, while the outer lateral tubercles are comparable in both, only the course of the fine ribs differing slightly. If Roemer's figure is accurate the stout whorl section suggests that it is a microconch.

TEXT-FIG. 22. *Placenticeras bidorsatum* (Roemer, 1841). GPIB 46, the original of Schlüter 1867, pl. 4, fig. 2; 1872*a*, pl. 15, fig. 6, paralectotype of *P. canaliculatum* Hyatt, 1903, Lower Campanian of Dülmen, Westphalia, German Federal Republic, × 1.

Of other European species, *P. bidorsatum* is easily separated from the Santonian *P. polyopsis* (see revision in Kennedy and Wright 1983) on details of tuberculation. The Coniacian *P. fritschi* de Grossouvre, 1894 (see revision in Kennedy 1984*a*, p. 38, pl. 2, fig. 1; pl 3, figs. 4 and 5; text-figs. 13C, 14C–E, 15) and *P. semiornatum* (d'Orbigny, 1850) (see revision in Kennedy 1984*a*, p. 44, pl. 1, figs. 4–6; pl. 4, figs. 3–5, 7, 8; text-figs. 13B and 29D) lack prominent tubercles when adult, as do the Santonian *P. paraplanum* Wiedmann, 1978 (p. 666, pl. 1, figs. 3 and 4; text-fig. 2A; see also Summesberger 1979, p. 152, pl. 13, figs. 53–57; text-figs. 38 and 39) and *P. maherndli* Summesberger, 1979 (p. 155, pl. 14, figs. 58–61; pl. 15, figs. 62–66; text-figs. 40–47). The evolutionary origin of *P. bidorsatum* probably lies in *paraplanum*, which shows rather similar ornament on the inner whorls (Summesberger 1979, fig. 55); the species transformation involving hypermorphosis to give the much larger *bidorsatum*.

TEXT-FIG. 23. *Placenticeras bidorsatum* (Roemer, 1841). GPIB unregistered *ex* Schlüter Collection, Lower Campanian of Broitzem, Braunschweig, German Federal Republic. This specimen is interpreted as an adult microconch, × 1.

Some authors (e.g. Brinkmann 1935; Wiedmann 1979) have referred to Hauer's *Ammonites milleri* (1866, p. 304, pl. 2, figs. 1–5), from Hemmerberg, between Kainach and Bärnbach, Steiermark, Austria, as a subspecies of *bidorsatum*. The species is probably Lower Campanian (Brinkmann records *Scaphites hippocrepis* var. *tenuis* Reeside from this area but the specimen is lost—*fide* Summesberger 1985, p. 161), but the type series, in the collections of the Geologische Bundesanstalt-Museum, Vienna is very poorly preserved indeed. Ornament consists of dense flexuous growth striae of variable strength, plus long ventrolateral clavi. In the best preserved specimen (text-fig. 11E) these number nine per half whorl and are opposite. They disappear on the body-chamber. If these small specimens are adult, the lack of flank tubercles immediately separates them from *bidorsatum*.

Occurrence. In Aquitaine most specimens are labelled P^1, a few P^{1-2} and one from Tauillard P^3. De Grossouvre (1894, 1901) shows it restricted to P^1. In Germany it occurs in the lower Lower Campanian of Dülmen, Westphalia and Braunschweig, Lower Saxony.

TEXT-FIG. 24. *Placenticeras bidorsatum* (Roemer, 1841). Copy of Roemer's original figures (1841, pl. 13, fig. 5) at their original size.

Genus HOPLITOPLACENTICERAS Paulcke, 1907, p. 183 (ICZN name no. 1348)
[= *Dechenoceras* Kayser, 1924, p. 175]

Type species. Hoplites–Placenticeras plasticus Paulcke, 1907, p. 186: ICZN Opinion 555, 1959: name no. 1629.

Diagnosis. Medium-sized, dimorphism demonstrated in some species. Moderately involute to moderately evolute, whorl section compressed to depressed, rectangular or trapezoidal. Ornament varies widely within and between species and through ontogeny. Basic ornament of umbilical, inner and outer ventrolateral tubercles linked by simple ribs may be modified by development of lautiform ribbing, outwards migration of umbilical bulla to inner lateral position with or without development of new umbilical bulla, when ribs may branch from mid-lateral tubercle, or ribs may be single to inner flank where they branch, with or without a tubercle. Inner and outer ventrolateral tubercles may be equal in number or one more numerous than the other. Both outer and inner ventrolaterals may link groups of ribs. Outer ventrolaterals may be equal or differentiate into long clavi linking groups of ribs, separated by single ribs with smaller clavi. They are generally opposite rather than alternate across venter. Clavi when well preserved are seen to be the bases of long septate spines (text-fig. 28A). Some species develop an outer lateral tubercle, others small ventral tubercles, equal to or more numerous than the outer ventrolateral. Ribbing may be augmented by fine lirae covering ribs and interspaces while in some fine ribbing or lirae dominate, with feeble umbilical bullae and minute ventrolateral tubercles or not.

In some coarsely ornamented species maturity is marked by a decline of all ornament leaving, ultimately, a body-chamber covered in dense lirae and striae only or with near-effaced tubercles.

Discussion. Study of the type species of Schlüter's diverse North German material (text-figs. 25–29) leads to a widening of the generic diagnosis to encompass a very variable group of ammonites. The nomenclature of European *Hoplitoplacenticeras* species is complicated by fact that Schlüter referred to the same specimens by different names on the four occasions he dealt with German representatives of the genus, while type designations made by certain revising authors are of questionable validity. In order to enable subsequent discussion I set out in chronological order the status of his species, and illustrate and designate lectotypes where appropriate.

1. *H. coesfeldiense* (Schlüter, 1867), p. 15, pl. 1, figs. 1–5. The lectotype, here designated, is GPIB 19*a*, a composite mould from Coesfeld, Westphalia, reillustrated here as text-fig. 27F. Mikhailov 1951, p. 82, refers to this specimen as the holotype which it is not, while this citation cannot be regarded as a valid lectotype designation under Article 74(a) of the 1964 *Code* then in operation, and Article 74(a) of the 1985 *Code* is ambiguous. Giers (1964, p. 275) and Atabekian and Khakhimov (1976, p. 86) refer to the original of Schlüter 1876, pl. 44, figs. 5 and 6 as the holotype; even if this is (as is possible) the same specimen as the original of Schlüter 1867, pl. 1, fig. 1, these are not valid lectotype designations for the same reasons, while if this specimen is *not* the same as the original of Schlüter 1867, pl. 1, fig. 1, it is not demonstrably part of the original type series. The originals of Schlüter 1867, pl. 1, figs. 2 and 3 are missing; they are paralectotypes of *H. coesfeldiense*, *H. striatocostatum* (Schlüter, 1872*a*) of which *H. vari* (Schlüter, 1872*b*) is the replacement name, and may be juvenile *vari* or *H. marroti* (Coquand, 1859). The original of Schlüter 1867, pl. 1, fig. 4, GPIB 19*d* is shown in text-fig. 27B, C; it appears to be within the limits of *H. coesfeldiense*. The original of Schlüter 1867, pl. 1, fig. 5, GPIB 19*b* is shown in text-fig. 28D; it is a paralectotype of *H. coesfeldiense*, showing features intermediate between the lectotype of that species and *H. dolbergense* (Schlüter, 1876).

2. *H. costulosum* (Schlüter, 1867), p. 17, pl. 2, figs. 1–4. The lectotype, here designated, is GPIB 20*c*, the original of Schlüter 1867, pl. 2, fig. 2, from Coesfeld, shown in text-fig. 29C, D. Mikhailov (1951, p. 83) refers to the original of pl. 2, fig. 3 as the holotype, which it is not, while this is not a valid lectotype designation for the reasons given above. Giers (1964, p. 279) refers to the original of Schlüter's pl. 2, fig. 2 as the ?lectotype, which cannot be construed as a valid lectotype designation under the *Rules* of whatever date. The original of Schlüter 1867, pl. 2, fig. 1, is a paralectotype of *H. costulosum* and lectotype of *H. striatocostatum* (Schlüter, 1872*a*) of which *vari* Schlüter 1872*b* (p. 92) is replacement name. It is GPIB 20*a*, is from Ahlten, Westphalia, and is shown in text-fig. 25. The original of Schlüter 1867, pl. 2, fig. 3 is GPIB 20*d*, a paralectotype of *costulosum*, shown in text-fig. 27D, E. The original of Schlüter 1867, pl. 2, fig. 4 is GPIB 20*b*, a paralectotype of *costulosum*, shown in text-fig. 28F, G.

3. *H. lemfoerdense* (Schlüter, 1872*b*) introduced as *nomen novum* for *Ammonites scaphitoides* Schlüter, 1872*a*, p. 63, pl. 19, figs. 1 and 2 *non* Coquand, 1853, p. 442, pl. 14, figs. 9 and 10. The holotype by monotypy ('Es hat sich bisher erst ein einziges Exemplar bei Haldem gezeit') is GPIB 50*a*, shown in text-fig. 29A, B.

4. *H. vari* (Schlüter, 1872*b*) (p. 92), was introduced as *nomen novum* for *A. striatocostatus* Schlüter 1872*a* (p. 65, pl. 20, figs. 1–4) *non* Meneghini, 1854, pl. 28. I here designate GPIB 20*a*, the original of Schlüter 1867, pl. 2, fig. 1, lectotype of the species. It is from Ahlten, Westphalia, and is shown in text-fig. 25.

5. *H. dolbergense* (Schlüter, 1876) (p. 159, pl. 44, figs. 1–4). Lectotype, here designated in the original of Schlüter 1876, pl. 44, figs. 2–3, ?4, GPIB 90*b*, from Dolberg, Westphalia, shown in text-fig. 28B, C, E. Giers (1964, p. 271) refers to this specimen as the holotype, which it is not. For reasons outlined above, this cannot be taken as a valid lectotype designation. The original of Schlüter 1876, pl. 44, fig. 1, GPIB 90*a* is a paralectotype of *dolbergense*, and is shown in text-fig. 28I, J.

The full resolution of the relationship between these species and which, if any, are synonyms is beyond the scope of the present work; the reader is referred to Giers (1964) for the most recent summary, although even this author refers the same specimens to *H. vari* and *H. costulosum* on successive pages!

Paulcke (1907) described a series of subgroups in his *Hoplites–Placenticeras plasticus*: *hauthali*, *crassus*, *costatus*, *semicostatus*, and *laeve*, all from Cerro Cazador, Patagonia. They represent what I consider a single variable species. The most compressed, involute, and feebly ornamented individuals (Paulcke 1907, pl. 14, fig. 4; pl. 15, figs. 2 and 3) have feeble to obsolete umbilical bullae, flexuous low ribs, and outer ventrolateral clavi only. In some (Paulcke 1907, pl. 15, fig. 3) the umbilical bulla strengthens with growth, and migrates out to an inner lateral position, leaving single ribs on the inner flank, with ribs branching from the migrated tubercle. As inflation and strength of ornament increases, juvenile ornament consists of umbilical bullae giving rise to one or two ribs with additional intercalatories, all with inner and outer ventrolateral tubercles (e.g. Paulcke 1907, pl. 10, fig. 3; pl. 12, fig. 3). As size increases, the umbilical bulla migrates outwards to a mid-lateral position in all large specimens (Paulcke 1907, pl. 10, figs. 1–3; pl. 11, figs. 1 and 2; pl. 12, figs. 1, 2, 4; pl. 13, figs. 1–3; pl. 14,

TEXT-FIG. 25. *Hoplitoplacenticeras* (*Hoplitoplacenticeras*) *vari* (Schlüter, 1872*b*). GPIB 20*a*, the lectotype, the original of Schlüter 1867, pl. 2, fig. 1, Upper Campanian of Ahlten, Westphalia, German Federal Republic, × 1.

leaving single ribs on the inner flank, and ribs branching at the migrated bulla, with additional intercalatories. This basic pattern of ornament is seen in *Hoplitoplacenticeras monju* Matsumoto, 1982 (p. 244, figs. 1 and 2; 1984*b*, p. 24, pl. 7, fig. 5; pl. 8, fig. 6), *H. fugen* Matsumoto, 1984*b* (p. 25, pl. 8, fig. 5). Species such as *H. marroti* (Coquand, 1859) (see p. 70, Pl. 2, figs. 3 and 4; Pl. 9, figs. 1–8, 11, 12; Pl. 10, figs. 1–12) has umbilical bullae giving rise to groups of flexuous ribs with inner and outer ventrolateral clavi like juvenile *plasticum*, but does not show outwards migration of umbilical bullae. *H. besairiei* Collignon, 1970 (p. 77, pl. 638, figs. 2349 and 2350) and *H. howarthi* Collignon, 1970 (p. 80, pl. 639, fig. 2351) are similar. *H. dolbergense* (Schlüter, 1876) (text-fig. 28A–E, H–J), the oldest of the North German species (*fide* Giers 1964) has early whorls with umbilical, inner and outer ventrolateral tubercles, much as in forms cited above, while the umbilical bulla migrates outwards in middle growth (text-fig. 29E) as in the *plasticum* group. It differs, however, in the development of looped ribbing at this stage. At maturity (text-fig. 29E) ornament departs from that of what I take to be adults of the *plasticum* group (e.g. Paulcke 1907, pl. 11, fig. 1) with the differentiation of the outer ventrolateral clavi into large ones linking groups of ribs separated by single ribs with small clavi only. In some specimens the larger clavi can be seen to be the bases of huge septate spines (text-fig. 28A). The adult body-chamber shows a near total loss of tubercles towards the aperture, leaving a shell covered by dense, fine lirae, striae, and riblets, a condition developed in the largest described *H. plasticum* (Paulcke 1907, pl. 11, fig. 1). The succeeding *H. coesfeldiense* (Schlüter, 1867) (text-fig. 27B, C, F) is little more than a finely ribbed version of *dolbergense*.

H. vari (Schlüter, 1872*b*) (text-fig. 25) has, when young, the simple ribbing and tuberculation of *marroti*. It differs at maturity in the appearance of dense lirae and striae covering all of the shell surface (text-figs. 25 and 26); as described below it may be a junior synonym of *marroti*, but good evidence is lacking. From this basic morphology there are passage forms where the ribbing declines and lirae dominate. Some of these co-occur with typical *vari* (e.g. text-figs. 26 and 27D, E) and there is a series of specimens that combine dense lirae with narrow, distant sinuous ribs, as in *H. costulosum* (Schlüter, 1867) (text-figs. 27A and 29C, D). Some juveniles of this species are ornamented by lirae only (text-figs. 27A and 28F, G) and adults referred to the species combine this with bullae and tiny ventral tubercles (text-figs. 26 and 27D, E).

These very different end members are not, so far as I can judge, separable into diagnosable genera or subgenera; some may be no more than intraspecific variants. *H. rejaudryi* (de Grossouvre, 1894) (p. 78, Pl. 7, figs. 1–5; Pl. 14, figs. 6 and 7) I also include in this group. It has common features with *H. coesfeldiense* and *dolbergense*, as I discuss below, as well as providing a possible ancestor to *H. lemfoerdense* (Schlüter, 1872*b*) (text-fig. 29A, B). This latter is the youngest of the North German species. It differs from all other Campanian species in the appearance of an additional set of ventral tubercles, so that there are mid-lateral, inner and outer ventrolateral, and ventral. This novel feature is also shown by the Upper Maastrichtian *H. lafresnayanum* (d'Orbigny, 1841) revised and misspelt *lasfresnayanum* in my recent account of the Calcaire à *Baculites* fauna (1986*a*, p. 48, pl. 9, figs. 4–6; text-fig. 10*b*), and I here separate them into a new subgenus:

H. (LEMFOERDICERAS) subgen. nov.

Type species. Ammonites lemfoerdensis Schlüter 1972*b*, p. 92; Upper Campanian, north Germany.

Diagnosis. Hoplitoplacenticeras with ventral tubercules on either side of siphonal line between the outer ventrolateral.

The remaining *Hoplitoplacenticeras* species are: *H. trangahyense* Collignon, 1970 (p. 76, pl. 638, fig. 2347), with lautiform ribbing, *H. antokazoense* Collignon, 1970 (p. 76, pl. 638, fig. 2348; pl. 639, fig. 2354), *H. gosseleti* (de Grossouvre, 1894) (p. 116, pl. 36, fig. 1), *H. rarecostatum* Khakhimov, 1976 (*in*

EXPLANATION OF PLATE 9

Figs. 1–8, 11, 12. *Hoplitoplacenticeras* (*Hoplitoplacenticeras*) *marroti* (Coquand, 1859). 1 and 2, SP unregistered, from Assise P^3, Leygonie, near Ribérac (Dordogne); 3 and 4, an adult microconch, SP unregistered *ex* Arnaud Collection, from Assise P^3, Petingaud, near Montmoreau (Charente), original of de Grossouvre 1894, pl. 9, fig. 3; 5 and 6, the holotype, EMP unregistered from the environs of Ribérac (Dordogne), note extensive restoration of outer whorl; 7, 8, 11 and 12, SP unregistered *ex* Arnaud Collection, from Assise P^2, Chavenat (Charente).

Figs. 9 and 10. *H.* (*H.*) *coesfeldiense* (Schlüter, 1867). BMNH C35981, from '$\frac{3}{4}$ m[ile] N of St Avit, NW of Le Buge on the Vezère River, Dordogne, France'.

All figs. × 1.

KENNEDY, *Hoplitoplacenticeras* (*Hoplitoplacenticeras*)

TEXT-FIG. 26. *Hoplitoplacenticeras* (*Hoplitoplacenticeras*) *vari* (Schlüter, 1872*b*). GPIB unregistered *ex* Schlüter Collection, Upper Campanian of Ennigerloh, Westphalia, German Federal Republic, × 1.

figs. 1 and 2) Atabekian and Khakhimov 1976, p. 88, pl. 10, fig. 5), and *H. vancouverense* (Meek, 1862, p. 317; see Usher 1952, p. 93, pl. 25, figs. 1 and 2; pl. 31, figs. 21 and 22).

Hassan (1971, p. 71) refers (?)*Scaphites kambysis* Zittel of Quaas 1902, p. 309, pl. 29, figs. 8–11 to *Hoplitoplacenticeras*. The inner whorls of this species are very similar to those of *Ammonites cunliffei* Forbes, 1846 (p. 109, pl. 8, fig. 2) from the Lower Maastrichtian Valudayur Group of Pondicherri, South India, type species by original designation of *Indoscaphites* Spath, 1953 (p. 14); the sutures are also closely comparable (Quaas 1902, pl. 29, fig. 11; Forbes 1846, pl. 8, fig. 2*d*). Whereas *cunliffei* has only umbilical and ventrolateral tubercles and these declining at maturity, *kambysis* develops, at the largest preserved diameters, outer ventrolaterals twice as numerous as the inner, to which they are linked by narrow ribs, then outer flank tubercles, then a row of tiny rounded siphonal tubercles which Quaas's reconstruction (1902, pl. 29, fig. 10) shows abruptly succeeded by a pair of tiny ventral tubercles that are more numerous than the outer ventrolateral. This last feature recalls *H.* (*Lemfoerdiceras*), but the remainder of the ornament is very different. I incline to the view that *kambysis* is a scaphitine homoeomorph of *Hoplitoplacenticeras*, given the other examples of convergence of Scaphitinae and Placenticeratidae (*Haresiceras* Reeside, 1927 and *Mancosiceras* Cobban, 1964; *Parrabinneyites* A. F. Leanza, 1964; *Discoscaphites* of *mandanensis* (Morton, 1842)

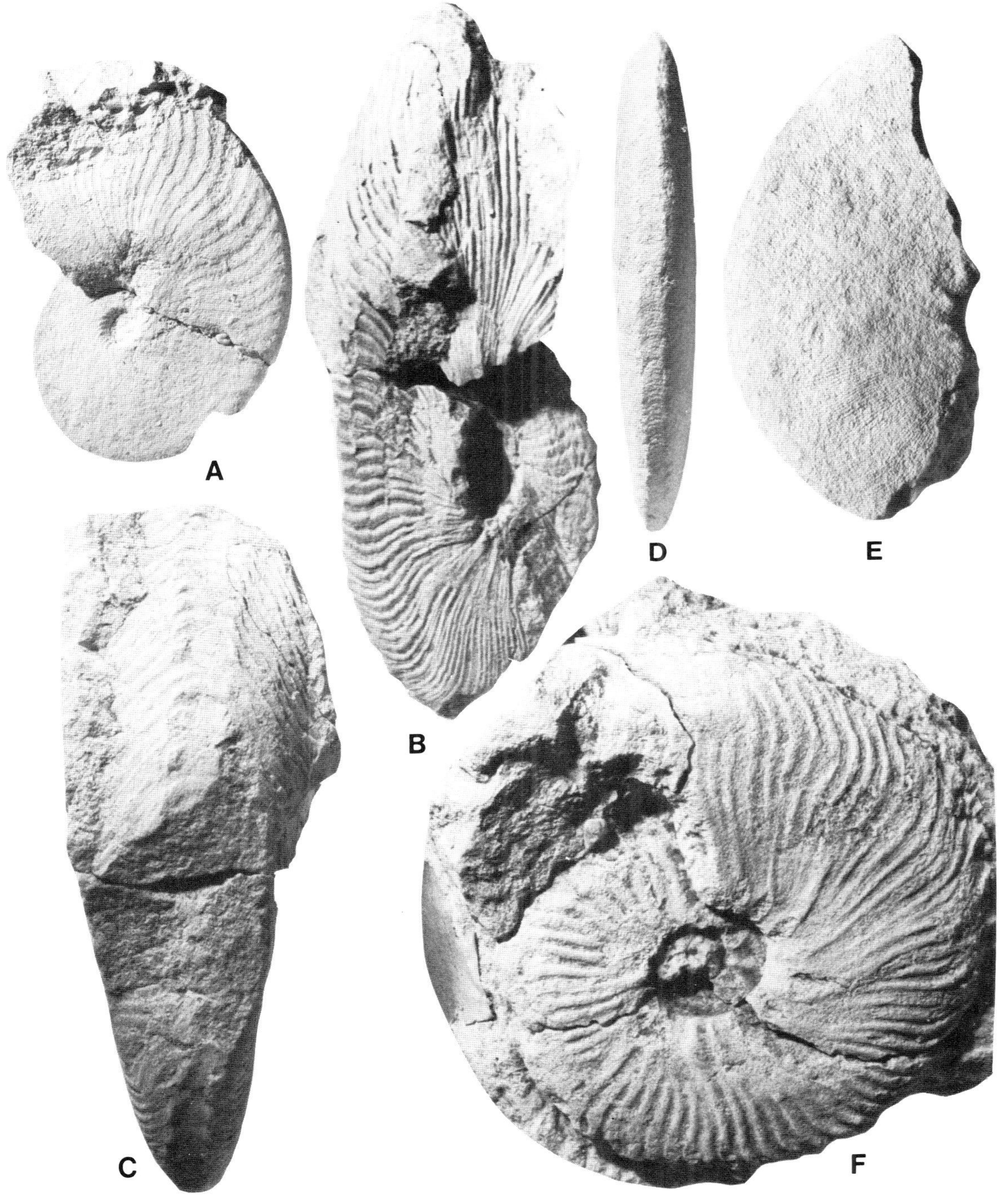

TEXT-FIG. 27. A, *Hoplitoplacenticeras* (*Hoplitoplacenticeras*) *costulosum* (Schlüter, 1867). GPIB 51, the original of Schlüter 1872*a*, pl. 20, figs. 5 and 6, Upper Campanian of Darup, Westphalia. B, C, F, *H.* (*H.*) *coesfeldiense* (Schlüter, 1867). B, C, GPIB 19*d*, a paralectotype, the original of Schlüter 1867, pl. 1, fig. 4, Upper Campanian of Coesfeld, Westphalia; F, the lectotype, GPIB 19*a*, from the same horizon and locality. D, E, *H.* (*H.*) cf. *costulosum* (Schlüter, 1867), GPIB 20*d*, the original of Schlüter 1867, pl. 3, fig. 3, a paralectotype of *costulosum*, Upper Campanian of Coesfeld, Westphalia. All specimens ×1.

type etc.). I suspect it belongs to the former, but having no material for study, cannot be certain. *Hoplitoplacenticeras awadi* Hassan, 1971 (p. 71, pl. 2, figs. 8 and 9) with siphonal tubercles also appears to be a scaphitid.

Occurrence. Upper Campanian to Upper Maastrichtian. Upper Campanian of Aude and northern Aquitaine in France, Netherlands, Sweden, Germany, European Russia, Soviet Central Asia, Caucasus, Crimea, Ukraine, Israel, Angola, Zululand, Madagascar, Japan, Texas and Wyoming in the USA, British Columbia in Canada, Patagonia. Upper Maastrichtian of Manche, France.

Hoplitoplacenticeras marroti (Coquand, 1859)

Plate 2, figs. 3 and 4; Plate 9, figs. 1–8, 11–12; Plate 10, figs. 1–12; Plate 12, figs. 1 and 2

1859 *Ammonites marroti* Coquand, p. 995.
1860 *Ammonites marroti* Coquand; Coquand, p. 158.
1867 *Ammonites coesfeldiensis* Schlüter, p. 14 (*pars*), pl. 1, figs. 2 and 3 only.
1872*a* *Ammonites striatocostatus* Schlüter, p. 65 (*pars*), *non* pl. 20, figs. 1–4.
1872*b* *Ammonites vari* Schlüter, p. 92 (*pars*).
1876 *Ammonites vari* Schlüter; Schlüter, p. 160 (*pars*).
1894 *Hoplites vari* Schlüter sp.; de Grossouvre, p. 118 (*pars*).
1894 *Hoplites vari* Schlüter var. *marroti* Coquand; de Grossouvre, p. 119, pl. 8, fig. 3; pl. 9, figs. 2 and 3.
1898 *Hoplites vari* Schlüter var. *marroti* Coquand; Choffat, p. 80, pl. 20, figs. 1–5 (*Puzosia marroti* in explanation to plate).
?1901 *Hoplites vari* Schlüter sp. var. nov. *praematura* Imkeller, p. 58, fig. 1.
1925 *Hoplitoplacenticeras vari* Schlüter; Diener, p. 178 (*pars*).
1925 *Hoplitoplacenticeras vari* (Schlüter) var. *Marroti* Coquand; Diener, p. 178.
1929 *Hoplites* cf. *vari* (Schlüter); Barrabé, p. 181, pl. 9, figs. 11–13.
1931 *Hoplites vari* Schlüter; Basse, p. 35, pl. 5, figs. 1–3; pl. 12, fig. 2; pl. 13, fig. 1.
1933 *Hoplitoplacenticeras* aff. *vari* Schlüter; Adkins, pp. 407 *et. seq.*
1947 *Hoplitoplacenticeras vari* (Schlüter); Chavan, p. 129, pl. 2, fig. 1.
1963 *Hoplitoplacenticeras marroti* (Coquand, 1859); Young, p. 63, pl. 2, figs. 5, 15, 17; pl. 17, figs. 3 and 4; pl. 20, figs. 2 and 3; pl. 21, figs. 1 and 4; pl. 81, fig. 4; text-figs. 9*b*, *c*, *f* and 11*a*.
non 1965 *Hoplitoplacenticeras* cf. *marroti* (Coquand); Howarth, p. 391, pl. 12, fig. 3; pl. 13, fig. 3.
1970 *Hoplitoplacenticeras marroti* (Coquand); Collignon, p. 75, pl. 638, figs. 2342 and 2343.
1976 *Hoplitoplacenticeras vari* (Schlüter); Atabekian and Khakhimov, p. 78, pl. 9, fig. 5; pl. 10, fig. 4; pl. 13, fig. 1.
1976 *Hoplitoplacenticeras marroti* (Coquand); Atabekian and Khakhimov, p. 81, pl. 9, fig. 6; pl. 13, figs. 2–4.
1984 *Hoplitoplacenticeras vari* (Schlüt.); Bilotte, pl. 37, figs. 2 and 3.
1985 *Hoplitoplacenticeras marroti* (Coquand, 1859); Kennedy, pl. 3, figs. 5–7.

Holotype. By monotypy, the specimen from Ribérac (Dordogne) (EMP unregistered) mentioned by Coquand (1859, p. 995) (Pl. 9, figs. 5 and 6).

Material. SP unregistered *ex* Arnaud Collection: 4 specimens labelled P^2, Chavenat (Charente); 1 specimen labelled P^3, Tauillard (Charente); 2 specimens labelled P^3, Bouteille (Dordogne), one the original of de

---

EXPLANATION OF PLATE 10

Figs. 1–12. *Hoplitoplacenticeras* (*Hoplitoplacenticeras*) *marroti* (Coquand, 1859). 1 and 2, SP unregistered *ex* Arnaud Collection, from Assise P^3, Bouteille (Dordogne), the original of de Grossouvre 1894, pl. 9, fig. 2; 3–5, 10, 11, Sinclair Collection, from Assise P^3, 1·5 km east of St Victor (Dordogne); 6 and 12, SP unregistered *ex* Toucas Collection, from Montmoreau (Charente); 7–9, LGSSP 118, from Aubeterre-sur-Dronne (Charente).
Figs. 13 and 14. *Eupachydiscus isculensis* (Redtenbacher, 1873). SP unregistered *ex* Arnaud Collection, from Assise P^1, Éraville (Charente). Original of de Grossouvre 1894, pl. 22, fig. 2.
Fig. 15. *Pachydiscus* (*Pachydiscus*) *launayi* de Grossouvre, 1894. SP unregistered *ex* Arnaud Collection, from Assise P^2, St-Hilaire-de-Montmoreau (Charente).
All figs. ×1.

PLATE 10

KENNEDY, *Hoplitoplacenticeras* (*Hoplitoplacenticeras*), *Eupachydiscus*, *Pachydiscus* (*Pachydiscus*)

Grossouvre 1894, pl. 9, fig. 2; 1 specimen labelled P^3, Petingaud [near Montmoreau] (Charente), the original of de Grossouvre 1894, pl. 9, fig. 3; 1 specimen labelled P^2, Nabinaud (Charente); 1 specimen labelled P^3, Courgeac-Beauvais (Charente); SP unregistered *ex* Toucas Collection from Montmoreau (Charente); MNHP, 1 specimen labelled Ribérac, St Paul Lizonne, P^{2-3} (Charente); LG55P 118, from Aubeterre-sur-Dronne (Charente); OUM KZ13950 and 13959 (counterpart) and numerous specimens in the Sinclair Collection (Allemans) from 1·5 km east of St Victor (Dordogne); SP unregistered from Leygonie near Ribérac (Dordogne).

Dimensions.

	D	*Wb*	*Wh*	*Wb*:*Wh*	*U*
SP Chavenat	54·9(100)	15·0(27·3)	28·3(51·5)	0·53	10·5(19·1)
Holotype	49·8(100)	15·8(31·7)	23·3(46·8)	0·68	— (—)
Sinclair Coll. 1	40·5(100)	15·2(37·5)	18·0(44·4)	0·84	8·2(20·2)
Sinclair Coll. 2	34·0(100)	10·8(31·8)	16·2(47·6)	0·67	8·6(25·3)

Description. All specimens are internal moulds. Coiling involute (U = 20–25 % of diameter), the small umbilicus of moderate depth with subvertical umbilical wall. Umbilical shoulder abruptly rounded; whorl section compressed (whorl breadth to height ratio 0·53–0·68) with greatest breadth low on flank intercostally and at umbilical bulla costally. Inner flanks broadly rounded, outer flanks flattened and convergent, truncated by ventrolateral facet in costal section; venter flattened. Ornament consists of twelve to fourteen umbilical bullae per whorl, small in most compressed but large and prominent in stoutest individuals. They are markedly prorsiradiate and give rise to prorsiradiate ribs which strengthen markedly across the flank. In general one rib is clearly linked to the bulla and a second (generally apertural) is tenuously linked or intercalates, while long intercalatories arise low on the flank. The ribs flex forwards to mid-flank and are convex, flexing back and concave on the outer flank before terminating at an inner ventrolateral tubercle, prominent in stout individuals (Pl. 10, figs. 1 and 2) but effacing markedly on the outer phragmocone whorls of compressed ones (Pl. 9, figs. 7 and 8). A broad rib extends from the inner ventrolateral tubercle to a prominent outer ventrolateral clavus; clavi are opposite rather than alternate and linked by a low, feeble broad rib. Total rib density ranges from sixteen to twenty per half whorl. Only one possible adult is present in the collection (Pl. 9, figs. 3 and 4). The umbilical seam egresses markedly, umbilical bullae decline, ribs crowd and weaken and there is a tendency to loop between umbilical and outer ventrolateral tubercles, the inner having declined. The ventral rib strengthens and provides a bar-like link between outer ventrolateral clavi. This specimen, an estimated 45 mm in diameter may be a microconch; other, incomplete specimens are septate to 60 mm (Pl. 9, figs. 7 and 8). None of the specimens show any trace of fine lirae or striae between or on the primary ribs, even when these are preserved on moulds of other molluscs in the same block (e.g. OUM KZ13950).

Suture line imperfectly exposed, with narrow saddles and broader, bifid lobes.

Discussion. The Aquitaine material forms a compact, only slightly variable assemblage, while material cited in the synonymy is closely comparable; the only exception is *Hoplitoplacenticeras marroti praematura* (Imkeller, 1901) (p. 58, fig. 1). As Howarth (1965, p. 392) notes, this form, from the Pattenauer Mergel of Stallauer Eck in the German Federal Republic, shows a very early loss of ventral tubercles, as in *H. rejaudryi* (De Grossouvre, 1894) (p. 146, pl. 8, figs. 4 and 5). It may be an adult microconch of the present species or some other. The majority of previous authors have regarded *marroti* of Coquand as a senior synonym of *vari* of Schlüter. This was introduced as a replacement name (Schlüter 1872*b*, p. 92) for *Ammonites striatocostatus* Schlüter, 1872*a* (p. 65, pl. 20, figs. 1–4), preoccupied by the combination *A. striatocostus* Meneghini (1854). *A. striatocostatus* of Schlüter 1872*a* (and hence *vari*) was itself based upon a series of specimens cited in 1872*a*, while the originals of *A. coesfeldiensis* Schlüter, 1867, p. 14 (*pars.*), pl. 1, figs. 2 and 3 and *A. costulosus* Schlüter, 1867, p. 17 (*pars.*), pl. 2, fig. 1 only were also cited and are syntypes. Giers (1964, p. 277) refers to the original of Schlüter 1872*a*, pl. 20, figs. 1 and 2, from Haldem, Westphalia (Geologisches und Paläontologisches Institut, Göttingen, no. 65-6) as the holotype, which it is not, and I have accordingly designated the original of Schlüter 1867, pl. 2, fig. 1 (text-fig. 25) lectotype above (p. 64). The two small specimens of Schlüter 1867 (pl. 1, figs. 2 and 3) may belong to *H. marroti*, although they have much straighter ribs in what may be idealized figures. In contrast the lectotype and other figured paralectotypes have a distinctive ornament of strong primary ribs and finer lirae that cover ribs and interspaces. Most of the French specimens are smaller than the types and other German specimens showing this ornament (e.g. Schlüter 1867, pl. 2, fig. 1; 1872, pl. 20, figs. 1–4; text-fig. 26), but at least one large fragment (OUM KZ3905/KZ3959) lacks all such ornament at a size comparable to this material. On this criterion I maintain *marroti* and *vari* as separate species, although admitting that additional work on German material may show that *vari* of certain authors are simply large *marroti*. Of other European species, *H. coesfeldiense* (Schlüter, 1867, p. 14, pl. 1, figs. 1 and 4; text-fig. 27B, C, F) has much finer, crowded ribs, 80 per whorl in the lectotype, in which all bear small inner ventrolateral tubercles and are linked periodically in groups of two or three to a very large outer ventrolateral clavus, in some cases separated by one to two ribs with outer ventrolateral clavi equal in size to the inner. *H. dolbergense* (Schlüter, 1876) (p. 159, pl. 44, figs. 1–4; 1872, p. 56, pl. 17, figs. 1–3) (as *coesfeldiensis*) has, at the same size as the present material (text-fig. 28A–E), ribs that loop between umbilical bullae and inner ventrolateral clavi, while the umbilical bulla migrates out to an inner lateral position. Schlüter's 1872*a* *Ammonites coesfeldiensis* (*non* 1867) (pl. 17, figs. 1–3) I take to be an adult macroconch. It shows a progressive weakening of ornament until, ultimately, the ribs become mere striae (text-fig. 29E, F). *H. costulosum* (Schlüter, 1867) (p. 17 (*pars*), pl. 2, figs. 2–4; 1872*a*, p. 60, pl. 20, figs. 5 and 6; text-fig. 27A) differs markedly from *H. marroti*. Juveniles are involute and smooth, ornamented by feeble bullae and lirae and striae only (text-figs. 27D, E and 28F, G); this ornament persists to a large size in some specimens (text-fig. 27D, E); in others the lirae are accompanied by fine, distant, flexuous low ribs. Some specimens bear clavi at the end of these ribs (text-fig. 29C, D) and are in certain respects transitional towards *H. vari* (text-fig. 26). *H. lemfoerdense* (Schlüter, 1872*b*) (p. 92 as *nomen novum* from *Ammonites scaphitoides* Schlüter, 1872*a*, p. 63, pl. 19, figs. 1 and 2 *non* Coquand, 1853; Schlüter 1876, p. 160, pl. 44, figs. 8 and 9 (on p. 160, Schlüter refers his pl. 44, figs. 1–4 to that species, manifestly an error)) has very fine ribs, mid-lateral tubercles, and small ventral tubercles more numerous than the outer ventrolateral clavi among other distinctive features. Similar fine ribbing characterizes *H. rejaudryi* (de Grossouvre, 1894), described further below, while the other French species, *H. gosseleti* (de Grossouvre, 1894) (p. 116, pl. 36, fig. 1) has more evolute coiling and simple straight narrow ribs with weak umbilical but strong inner and outer ventrolateral tubercles.

Some of the variable *H. plasticum* of Paulcke (1907), notably *H. p. costatum* Paulcke, 1907 (p. 34, pl. 11, fig. 2; pl. 12, figs. 1–3) have been compared to the present species. If, as seems likely, all of the Patagonian material represents no more than a single variable species, the two are very different, with many forms of *plasticum* showing the umbilical tubercle migrating out to a mid-lateral position, more evolute coiling, and an inflated whorl section. Such morphologies are not seen in the Aquitaine assemblage, although individual specimens are similar, if more strongly tuberculate at comparable whorl inflation in some cases, with ribs that are straight rather than flexuous.

Occurrence. Museum specimens are labelled as from P^2–P^3 in Aquitaine; according to de Grossouvre (1894, 1901) it is restricted to P^3, which is my experience also. The species occurs at a comparable lower Upper Campanian horizon in Ariège, France (Bilotte 1984, pl. 37, figs. 2 and 3), possibly Coesfeld in Westphalia, Vijlen near Cotessen on the Belgian–Netherlands border, Portugal, Israel, Soviet Central Asia, Madagascar, and Texas, USA.

Hoplitoplacenticeras coesfeldiense (Schlüter, 1867)

Plate 9, figs. 9 and 10; text-fig. 27B, C, F

1867 *Ammonites Coesfeldiensis* Schlüter, p. 14 (*pars*), pl. 1, figs. 1 and 4 only (*non* 2, 3, ? = *H. marroti* (Coquand, 1859); *non* fig. 5? = *H. dolbergense* (Schlüter, 1876)).

1872*a* *Ammonites Coesfeldiensis* Schlüter; Schlüter, p. 56 (*pars*), *non* pl 17, figs. 1–3 (= *H. dolbergense* (Schlüter, 1876)).

1876 *Ammonites Coesfeldiensis* Schlüter; Schlüter, p. 159, pl. 44, figs. 5–7.

1894 *Hoplites coesfeldiensis* Schlüter sp.; de Grossouvre, p. 120.

1907 *Hoplitoplacenticeras coesfeldiensis*; Paulcke, p. 214.

1925 *Hoplitoplacenticeras coesfeldiense* Schlüter; Diener, p. 177.

1936 *Hoplites coesfeldiensis* Schlüter; Nalivajko, p. 33, pl. 15, fig. 36.

non 1938 *Hoplitoplacenticeras coesfeldiense* Schlüter; Roman, p. 506, pl. 53, fig. 492 (= *H. dolbergense* Schlüter, 1876).

1951 *Hoplitoplacenticeras coesfeldiense* Schlüter; Mikhailov, p. 81, pl. 13, figs. 53 and 54; pl. 15, fig. 62; pl. 18, fig. 89.
non 1951 *Hoplitoplacenticeras coesfeldiense* Schlüt. var. *schlüteri* nom. nov.; Mikhailov, p. 82 (*pars*), pl. 15, figs. 60 and 61.
non 1951 *Hoplitoplacenticeras coesfeldiense* Schlüt. var. *costulosa* Schlüter; Mikhailov, p. 83, pl. 14, fig. 57 (= *H. costulosum* (Schlüter, 1867).
1953 *Dechenoceras coesfeldiense* Schlüter; Ødum, p. 23, pl. 3, fig. 2.
1959 *Hoplitoplacenticeras coesfeldiense* (Schlüter); Naidin and Shimanskij, p. 193, pl. 15, fig. 1.
non 1959 *Hoplitoplacenticeras vari* var. *schlüteri* Michailov; Naidin and Shimanskij, p. 193, pl. 15, fig. 2.
1964 *Hoplitoplacenticeras coesfeldiense* (Schlüt.); Giers, p. 275, pl. 7, figs. 2 and 3.
?*non* 1973*a* *Hoplitoplacenticeras* cf. *H. coesfeldiense* (Schlüter) var. *schlüteri* Mikhailov; Cobban, p. C60, fig. 76.1 (? = *H. dolbergense* (Schlüter, 1876)).
1974 *Hoplitoplacenticeras coesfeldiense coesfeldiense* (Schlüter, 1867); Naidin, p. 187, pl. 69, fig. 3; pl. 70, fig. 1.
non 1974 *Hoplitoplacenticeras coesfeldiense costulosum* (Schlüter, 1867); Naidin, p. 187, pl. 64, fig. 4 (= *H. costulosum* Schlüter, 1867).
1976 *Hoplitoplacenticeras coesfeldiense* (Schlüter); Atabekian and Khakhimov, p. 86, pl. 13, fig. 5.

Types. Lectotype, herein designated (p. 64), GPIB 19*a*, the original of Schlüter 1867, p. 15, pl. 1, fig. 1, from Coesfeld, Westphalia, reillustrated here as text-fig. 27F. Of the paralectotypes, the originals of Schlüter's pl. 1, figs. 2 and 3 have not been traced and are either juvenile *Hoplitoplacenticeras vari* (Schlüter, 1872*b*) or *H. marroti* (Coquand, 1859); the original of Schlüter's 1867, pl. 1, fig. 4 is GPIB 19*d* (text-fig. 27B, C) and belongs to the present species; the original of Schlüter's 1867, pl. 1, fig. 5, GPIB 19*b* (text-fig. 28D) seems closer to *H. dolbergense* (Schlüter, 1876).

Material. BMNH C35981 from '3/4 m[ile] N of St. Avit, N.W. of Le Buge on the Vezère River, Dordogne, France'.

Description. The lectotype (text-fig. 27F) is a crushed composite mould. Coiling involute, 80% of previous whorl covered. Umbilicus shallow, small, comprising approximately 23% of diameter. Primary ribs arise at umbilical seam and develop into small umbilical bullae, prominent on inner whorl but declining and migrating outwards on (damaged) outer whorl. They give rise to groups of 2–3 flexuous prorsiradiate primary ribs; shorter intercalated ribs arise on mid- and outer flank to give estimated total of 80 ribs per whorl. All ribs bear a small inner ventrolateral tubercle on the first two thirds of the outer whorl. Large, elongate outer ventrolateral clavi link groups of 2–3 ribs and are commonly radially disposed from umbilical bullae. There are sometimes 2–3 ribs with small outer ventrolateral clavi (more-or-less equal to the inner ventrolateral) between these very large clavi. Flank ornament weakens on last half whorl, a feeble inner lateral tubercle appears on two ribs, while towards the aperture ribs weaken and decline markedly in strength, inner ventrolateral clavi decline, outer ventrolaterals become smaller. At several points the matrix shows that the outer ventrolateral clavi are the bases of large spines.

Paralectotype GPIB19*b* (Schlüter 1867, pl. 1, fig. 4) (text-fig. 27B, C) is crushed and distorted, but shows a rib strength and density like that of the lectotype. It has sharp umbilical bullae throughout, but only tiny equal inner and outer ventrolateral tubercles at the smallest diameter visible.

The French specimen matches the lectotype closely, has an estimated 38 ribs per half whorl, feeble inner ventrolateral tubercles that efface as size increases with large outer ventrolateral clavi linking groups of 2 or 3 ribs separated by 2–3 ribs with small clavi, not linked into groups.

TEXT-FIG. 28. A–E, H–J, *Hoplitoplacenticeras* (*Hoplitoplacenticeras*) *dolbergense* (Schlüter, 1876). A, GPIB unregistered, from Ennigerloh, Westphalia (*ex* Schlüter Collection); note preservation of septate spines in entirety (arrow); B, C, E, GPIB 90*b*, the lectotype, the original of Schlüter 1876, pl. 44, figs. 2, 3 ?4, Upper Campanian of Dolberg, Westphalia; D, GPIB 19, the original of Schlüter 1867, pl. 1, fig. 5, a paralectotype of *H.* (*H.*) *coesfeldiense* and transitional in some respects between *coesfeldiense* and *dolbergense*, Upper Campanian of Coesfeld, Westphalia; H, GPIB 19C, the original of Schlüter 1872*a*, pl. 17, fig. 3, a paralectotype from the same horizon and locality as the preceding; I, J, GPIB 90*a*, the original of Schlüter 1876, pl. 44, fig. 1, a paralectotype, Upper Campanian of Darup, Westphalia. F, G, *H.* (*H.*) cf. *costulosum* (Schlüter, 1867), GPIB 20*b*, a paralectotype of *costulosum* and the original of Schlüter 1867, pl. 2, fig. 4, Upper Campanian of Coesfeld, Westphalia. All specimens × 1.

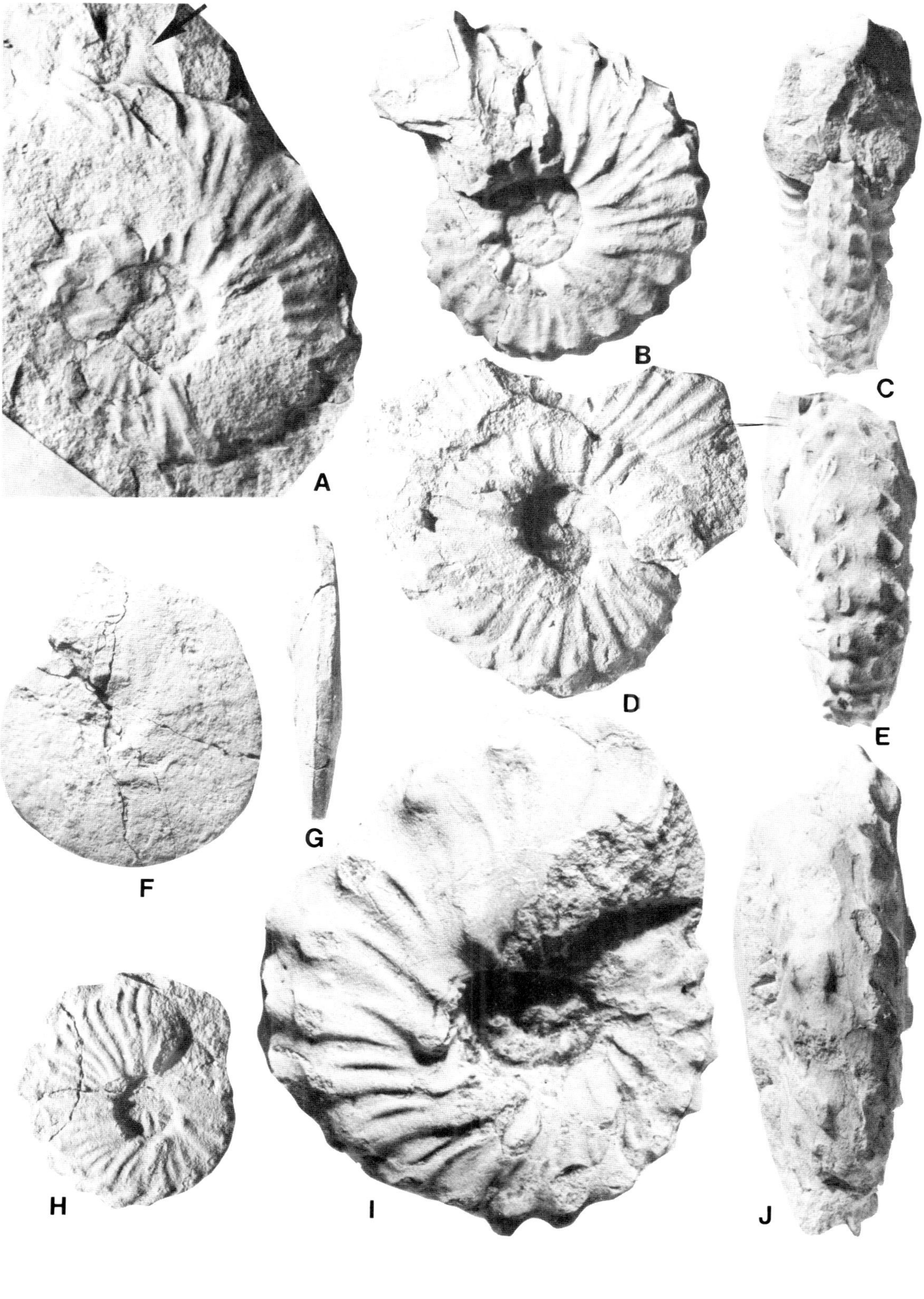
A
B
C
D
E
F
G
H
I
J

Discussion. The French specimen clearly belongs to *H. coesfeldiense*. The German paralectotype shows certain differences of ventrolateral tuberculation, but until there is a full revision of the abundant German material I can do no more than regard it as conspecific, as have previous authors. As noted above, the two missing paralectotypes (Schlüter 1867, pl. 1, figs. 2 and 3) are juveniles of either *H. vari* or *H. marroti*. The original of Schlüter 1867, pl. 1, fig. 5 (text-fig. 28D) is much coarser ribbed and seems to be closer to *H. dolbergense* (Schlüter, 1876) (p. 159, pl. 44, figs. 1–4), the lectotype of which is shown in text-fig. 28B, C, E. This has much coarser, strikingly lautiform ornament, ribs looping between coarse umbilical and inner ventrolateral tubercles that are equal in number to the outer ventrolateral, which are all of the same size in the early growth stages, while the umbilical bullae migrate out to an inner lateral position so that there are simple ribs on the inner flank, with groups of ribs arising from an inner lateral position. *H. coesfeldiense schlüteri* Mikhailov, 1951 (p. 82, pl. 15, figs. 60 and 61) to which *Ammonites coesfeldiensis* of Schlüter 1872, pl. 17, figs. 1–3 has been referred, seems to be an adult macroconch *dolbergense* (text-fig. 29E, F). This specimen shows inner whorls identical to those of the lectotype of *dolbergense* (text-fig. 28B, C, E). There are thus nine to ten bullate primary ribs with a well-developed inner lateral tubercle. Between are one or two non-tuberculate primaries of variable length. Ribs increase by branching and intercalation, are straight and prorsiradiate. They connect in pairs to strong inner ventrolateral clavi or not, while there are aperturally displaced outer ventrolateral clavi that are alternate rather than opposite on the venter. Beyond 85 mm, ornament changes progressively; umbilical and inner lateral tubercles decline, while small outer ventrolateral clavi appear between the larger, so that the total number of outer ventrolaterals exceeds that of the inner, which rapidly disappear. Ribs weaken to mere riblets and lirae, some periodically strengthened, with feeble inner and outer lateral tubercles.

Whether *dolbergense* and *coesfeldiense* are extremes of one variable species or fully distinct cannot be determined without study of the abundant German material; note that Giers (1964, p. 275) refers to the latter as a 'mutations-form' of the former.

H. coesfeldiense is readily separated from species such as *H. marroti*, *vari*, and *costulosum* by its distinctive pattern of ribbing and tuberculation. There are some similarities to *H.* (*Lemfoerdiceras*) *lemfoerdense* (Schlüter, 1872*b*) (text-fig. 29A, B), but this remarkable species develops an additional ventral row of tubercles that are more numerous than the outer ventrolateral. *H. rejaudryi* (de Grossouvre, 1894) (p. 146, pl. 8, figs. 2 and 4) is a further finely ribbed form differing from *H. coesfeldiense* in having primary ribs extending to mid-flank, where they bi- and trifurcate, and lacking the prominent looping of ribs at large outer ventral clavi.

Occurrence. The Aquitaine specimen is imprecisely localized within the Upper Campanian. In Germany the species occurs in the lower Upper Campanian according to Giers (1964); it is also recorded from the Upper Campanian of European Russia and Soviet Central Asia. There are two fragments in the Naturhistorische Museum, Maastricht, from the Upper Campanian Zeven Wegen Chalk at Vijlenerbosch, Limburg, the Netherlands (NMM GK840 *ex* Felder Collection).

Hoplitoplacenticeras dolbergense (Schlüter, 1876)

Plate 11, figs. 1 and 2

1867 *Ammonites Coesfeldiensis* Schlüter, p. 14, pl. 1, fig. 5 only.
1872*a* *Ammonites Coesfeldiensis* Schlüter; Schlüter, p. 56 (*pars*), pl. 17, figs. 1–3.
1876 *Ammonites Dolbergensis* Schlüter, p. 159, pl. 44, figs. 1–4.

TEXT-FIG. 29. A, B, *Hoplitoplacenticeras* (*Lemfoerdiceras*) *lemfoerdense* (Schlüter, 1872*b*), the holotype, by monotypy, GPIB 50*a*, the original of Schlüter 1872*a*, pl. 19, figs. 1 and 2, Upper Campanian of Haldem, Westphalia. C, D, *H.* (*Hoplitoplacenticeras*) *costulosum* (Schlüter, 1867), GPIB 20*c*, the lectotype, the original of Schlüter 1867, pl. 2, fig. 3, Upper Campanian of Coesfeld, Westphalia. E, F, *H.* (*H.*) *dolbergense* (Schlüter, 1876), GPIB 19*e*, the original of Schlüter 1872*b*, pl. 17, figs. 1 and 2, a paralectotype, Upper Campanian between Beckum and Ennigerloh, Westphalia. All specimens ×1.

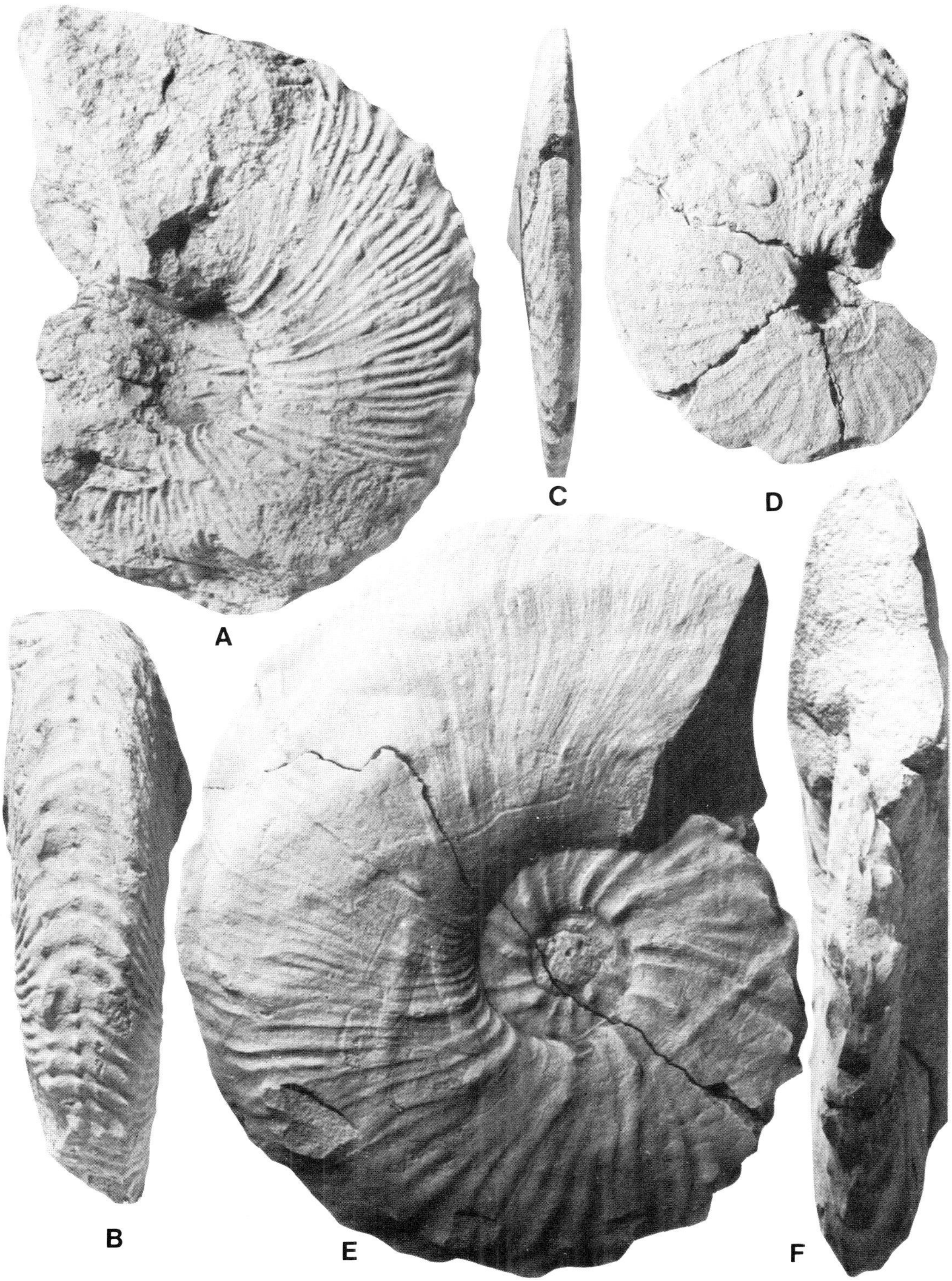
A
B
C
D
E
F

1894 *Hoplites dolbergensis* Schlüter; de Grossouvre, p. 119.
1938 *Hoplitoplacenticeras dolbergense* Schlüter; Roman, p. 506.
1938 *Hoplitoplacenticeras coesfeldiense* Schlüter; Roman, p. 506, pl. 53, figs. 492, 492*a*; 492 on p. 508.
1951 *Hoplitoplacenticeras coesfeldiense* Schlüt. var. *schlüteri* nom. nov.; Mikhailov, p. 82 (*pars*), pl. 15, figs. 60 and 61.
1959 *Hoplitoplacenticeras vari* var. *schlüteri* Mikhailov; Naidin and Shimanskij, p. 193, pl. 15, fig. 2.
1964 *Hoplitoplacenticeras dolbergense* (Schlüt.); Giers, p. 271, pl. 6, figs. 2–7; pl. 7, fig. 1; text-fig. 7.
1973*a* *Hoplitoplacenticeras* cf. *H. coesfeldiense* var. *schlüteri* Mikhailov; Cobban, p. C60, fig. 76.1.
1974 *Hoplitoplacenticeras coesfeldiense coesfeldiense* (Schlüter, 1867); Naidin, p. 187 (*pars*), pl. 69, fig. 3 only.
1975 *Hoplitoplacenticeras dolbergense* (Schlüter, 1876); Schmid and Ernst, p. 346.

Types. Lectotype herein designated (p. 64) GPIB 90*b*, the original of Schlüter 1876, pl. 44, figs. 2 and 3, ?4 (text-fig. 28B, C, E) from the Upper Campanian of Dolberg, Westphalia. Paralectotype GPIB 90, the original of Schlüter 1876, pl. 44, fig. 1 (text-fig. 28I, J) is from the same horizon at Darup, Westphalia.

Material. SP unregistered *ex* Arnaud Collection, from Assise P^3 at Petingaud (Charente).

Description. The specimen is a composite mould of the last part of the body-chamber of a specimen with an original diameter of approximately 65–70 mm. At the smallest diameter visible there are strong outer ventrolateral clavi present, but for most of the specimen ornament consists of umbilical bullae giving rise to bunches of fine lirae and riblets, while similar ornament arises at the umbilical shoulder between bullae so that the whole flank is covered in this fine flexuous ornament, which sweeps over the rounded ventrolateral shoulders to cross the venter in a linguoid peak.

Discussion. This curious fragment is referred to *Hoplitoplacenticeras dolbergense* because of its distinctive adult ornament, thus far only described in that species. Schlüter (1872*a*, pl. 17, figs. 1 and 2, as *coesfeldiense*) figured a larger specimen some 120 mm in diameter, while Giers (1964, pl. 7, fig. 1) illustrated a specimen showing the same abrupt change to adult ornament only 75 mm approximately in diameter. I take the former to be an adult macroconch, the latter, and the present specimen to be microconchs.

Occurrence. Assise P^3 in Aquitaine, lower Upper Campanian in Germany, and European Russia. A further possible occurrence is in the *Baculites asperiformis* Zone Steele Shale of Carbon County, Wyoming.

Hoplitoplacenticeras rejaudryi (de Grossouvre, 1894)

Plate 7, figs. 1–5; Plate 14, figs. 6 and 7

1894 *Sonneratia rejaudryi* de Grossouvre, p. 146, pl. 8, fig. 4.
1894 *Sonneratia rara* de Grossouvre, p. 148, pl. 8, fig. 2.

Types. Holotype, by original designation, SP unregistered *ex* Rejaudry Collection, from Assise P^3, Saint-Médard-de-Barbezieux (Charente); paratype, SP unregistered *ex* Arnaud Collection, from the same horizon and locality. The holotype of *Sonneratia rara*, SP unregistered *ex* Arnaud Collection is from the same horizon and locality.

EXPLANATION OF PLATE 11

Figs. 1 and 2. *Hoplitoplacenticeras* (*Hoplitoplacenticeras*) *dolbergense* (Schlüter, 1876). SP unregistered *ex* Arnaud Collection, from Assise P^3, Petingaud (Charente).
Figs. 3 and 4. *Menabites* (*Delawarella*) *delawarensis* (Morton, 1830). SP unregistered *ex* Arnaud Collection, the paralectotype of *Mortoniceras campaniensis* de Grossouvre, 1894, pl. 13, fig. 1, labelled Larzac (Dordogne), but said by de Grossouvre to be from St-Hilaire-de-Montmoreau (Charente).
Figs. 5 and 6. *Praemuniericeras* sp. MNHP B17226, from Assise P, La Vaure, near Montignac (Dordogne).
Figs. 7 and 8. *Menabites* (*Australiella*) cf. *besairiei* (Collignon, 1938). SP unregistered *ex* Arnaud Collection, a paralectotype of *Mortoniceras campaniensis* de Grossouvre, 1894, from Assise P^{1-2}, Montmoreau (Charente).
All figs. ×1.

KENNEDY, *Hoplitoplacenticeras* (*Hoplitoplacenticeras*), *Menabites* (*Delawarella*), *Praemuniericeras*, *Menabites* (*Australiella*)

Dimensions.

	D	Wb	Wh	Wb:Wh	U
Holotype	79·5(100)	—(—)	34·5(43·3)	—	21·0(26·4)

Description. Coiling fairly involute, 70% of previous whorl covered. Umbilicus of moderate width; shallow, with flattened, outwards-inclined umbilical wall; umbilical shoulder narrowly rounded. Whorl section compressed (whorl breadth to height ratio 0·78), with greatest breadth at umbilical shoulder. Inner flanks broadly rounded; outer flanks flattened and convergent; ventrolateral shoulders broadly rounded; venter flattened. The early growth stages are shown by the holotype of *S. rara* (Pl. 7, figs. 4 and 5), a composite internal mould with a maximum whorl height of 9 mm. Narrow crowded ribs arise at the umbilical seam and pass up the umbilical wall, where they are feebly concave. They are markedly prorsiradiate on the inner flank and without an umbilical bulla. Primaries bifurcate at mid-flank and concave secondaries extend to the ventrolateral shoulder, as do one or two intercalatories inserted at mid-flank, so that eight primaries correspond to twenty-six secondaries on the fragment. All ribs terminate in a tiny outer ventrolateral clavus, linked to the opposing clavus over the flattened venter by a broadly convex rib.

The much larger holotype of *rejaudryi* has an estimated thirty narrow primary ribs arising at the umbilicus, concave on the umbilical wall and straight and prorsiradiate to mid-flank, lacking an umbilical bulla as in the paratype. They strengthen into an incipient mid-lateral bulla, from which two or three secondaries arise, while one, sometimes two intercalatories arise at the same point. They are straight or feebly concave to the ventrolateral shoulder, where they project forwards and cross the venter in a broad convexity. There are no obvious ventrolateral tubercles at any stage in development visible up to 68 mm diameter. The last quarter whorl shows a weakening of flank ornament, possibly suggesting the onset of maturity, while ribbing weakens over the siphonal region, and the ends of the ribs thicken into feeble obliquely placed clavi.

The third specimen, although much distorted, shows comparable fine, wiry ribbing as the holotype, but differs in showing a feebly concave venter throughout, with a trace of an outer ventrolateral tubercle on each side of the concavity, while there are occasional large outer ventrolateral tubercles linking pairs of ribs.

Discussion. De Grossouvre (1894) referred this species to the Lower Albian *Sonneratia* Bayle, 1878, and Diener (1925, p. 179) to *Parapachydiscus* Hyatt, 1900, a synonym of *Pachydiscus* Zittel, 1884. It most closely resembles, however, *H.* (*Lemfoerdiceras*) *lemfoerdense* (Schlüter, 1972*b*) (p. 92) (as *nomen novum* for *scaphitoides* Schlüter, 1872*a*, p. 63, pl. 19, figs. 1 and 2) (text-fig. 29A, B) and *H. dolbergense* (Schlüter, 1876) (p. 159, pl. 44, figs. 1–4) (text-fig. 29E, F). All show a primary rib extending to mid-flank at some stage in development, while *lemfoerdense* has very similar wiry flank ribs that subdivide at a lateral tubercle, incipient in the present species. They differ in the presence of small inner ventrolateral tubercles, large outer ventrolaterals linking groups of ribs and smaller, more numerous ventral tubercles in *lemfoerdense*. There are also comparisons with the highly distorted paralectotype of *H. coesfeldiense* (Schlüter, 1867) (see above; text-fig. 27B, C) which is, in terms of ribbing and ventral tuberculation, intermediate between the two species as defined by their types in that it lacks the long clavi linking groups of ribs separated by ribs with smaller clavi seen in the lectotype of *coesfeldiense* (text-fig. 27F).

Occurrence. As for types.

Superfamily ACANTHOCERATACEAE de Grossouvre, 1894, p. 22

[*nom. correct.* Wright and Wright 1951, p. 24 (*pro* Acanthoceratida Hyatt, 1900, p. 585), *nom. transl. ex* Acanthoceratidae Hyatt, 1900, p. 585, *nom. correct. ex* Acanthoceratidés de Grossouvre, 1894]

Family COLLIGNONICERATIDAE Wright and Wright, 1951, p. 30

[*nom. subst. pro.* Prionotropidae Zittel, 1895, p. 430 (*ex Prionotropis* Meek, 1876*a*, p. 453, *non* Fieber, 1853, p. 127; = *Collignoniceras* Breistroffer, 1947 (unpaged) (= Prionocyclidae Breistroffer, 1947 (unpaged) (*ex Prionocyclus* Meek, 1876*a*, p. 298, ineligible as family type)]

Subfamily TEXANITINAE Collignon, 1948, p. 54(9)

[*nom. transl.* Wright, 1957, p. L429 *ex* Texanitidae Collignon, 1948]

Genus SUBMORTONICERAS Spath, 1926, p. 79

[= *Butticeras* Anderson, 1958, p. 272; *Antsirasirella* Collignon, 1969, p. 201; *Ankilizatella* Collignon, 1971, p. 50]

Type species. Mortoniceras woodsi Spath, 1921, p. 232, pl. 21, fig. 1, by original designation.

Submortoniceras sp.

Plate 6, figs. 3 and 4

Material. SP unregistered. Unlocalized, but by its preservation from Assise P of northern Aquitaine.

Description. The specimen is highly distorted and a composite internal mould. The original diameter was approximately 65 mm. Coiling appears to have been relatively involute. The whorl section is badly distorted, but the flanks were flattened and convergent, with a flattened venter. Primary ribs arise from strong bullae (1) that project into the umbilicus. They may arise singly or a second rib be tenuously linked to the bulla; a few intercalate low on the flank. All ribs bear a feeble inner lateral tubercle (2), stronger submarginal (3) and marginal (4), with the external (5) strongest of all. There is a somewhat worn siphonal keel.

Discussion. This specimen at first sight resembles the *Menabites* (*Delawarella*) from Aquitaine, but the equal number of tubercles in all rows plus coiling show it to be a *Submortoniceras*, not previously known from the region. It is too small and distorted for specific determination. *Submortoniceras* is very rare indeed in Europe; the only other specimens I know of are from the Lower Campanian of northern Spain (e.g. Ernst Collection).

Occurrence. As for material.

Genus MENABITES Collignon, 1948, p. 64 (19)

Type species. Menabites menabensis Collignon, 1948, p. 7 (64), pl. 17, figs. 3 and 4; pl. 18, fig. 1, by subsequent designation by Wright 1957 (p. L432).

Subgenus DELAWARELLA Collignon, 1948, p. 64 (19)

Type species. Ammonites delawarensis Morton, 1830, p. 244, pl. 2, fig. 4, by original designation.

Menabites (*Delawarella*) *delawarensis* (Morton, 1830)

Plate 11, figs. 3 and 4; text-fig. 30

1830 *Ammonites delawarensis* Morton, p. 244, pl. 2, fig. 4.
1894 *Mortoniceras campaniense* de Grossouvre, p. 84 (*pars*), pl. 13, fig. 1 only.
1948 *Menabites* (*Delawarella*) *delawarensis* Morton; Collignon, p. 29 (86) (with full synonymy).
1962 *Menabites* (*Delawarella*) *delawarensis* (Morton); Reeside, p. 131, pl. 72, figs. 1–3; pl. 73, figs. 1–5; pl. 74, fig. 2.
1963 *Delawarella delawarensis* (Morton, 1830); Young, p. 111, pl. 55, fig. 5; pl. 61, figs. 1–6; pl. 63, fig. 2; text-figs. 15*e*, 20*d*, 25*b*, *f*, *g*, 27*c*, 29*f*.
1980 *Menabites* (*Delawarella*) *delawarensis* (Morton, 1830); Klinger and Kennedy, p. 320, figs. 244–247A.
1985 *Delawarella campaniensis* (de Grossouvre, 1894); Kennedy, pl. 3, figs. 3 and 4.

Types. The original of Morton 1830, pl. 2, fig. 4 is the only figured syntype. Reeside (1962, p. 132) says this specimen is probably lost. Young (1963, pl. 61, fig. 3) figures what he states to be the holotype, but this is no. 1949 in the Collections of the Academy of Natural Sciences, Philadelphia, also illustrated by Whitfield (1892, pl. 42, figs. 6 and 7) and Reeside (1962, pl. 72, figs. 1–3). It bears no resemblance whatsoever to the figured type and this observation is incorrect.

Material. SP unregistered *ex* Arnaud Collection, from Assise P[1]. The associated labels state it to be from Larzac (Dordogne); de Grossouvre (1894, p. 85; explanation of pl. 13, fig. 1) states it to be from P[1] at Saint Hilaire-de-Montmoreau (Charente). It is a syntype of *Mortoniceras campaniense* de Grossouvre, 1894. A second specimen in the Sinclair Collection (Allemans, Dordogne) is from a comparable horizon at Larmalice, Allemans (Dordogne).

TEXT-FIG. 30. *Menabites* (*Delawarella*) *delawarensis* (Morton, 1830). Sinclair Collection, upper Lower Campanian, Larmalice (Allemans) (Dordogne), × 1.

Description. Moderately evolute, with umbilicus comprising approximately 30% of diameter, broad, with the submarginal tubercles housed in notches in the wall of the succeeding whorl. Whorl section slightly depressed (whorl breadth to height ratio 1·02 in costal section) with greatest breadth at umbilical bulla. There are thirteen primary ribs per half whorl. They are broad, distant, blunt, straight, and prorsiradiate. All bear a prominent umbilical bulla (1), rounded inner lateral tubercle (2), and much stronger, conical submarginal tubercle (3). At the smallest size visible (Pl. 11, figs. 3 and 4) the ribs bifurcate at this tubercle, terminating at blunt external clavi (5), but for most of the development stages visible there is an additional, marginal tubercle. As size increases the tubercles become more equally developed, although the submarginal is the most prominent (text-fig. 30).

Suture poorly exposed, with little-incised, bifid elements.

Discussion. I have matched the French material with casts of specimens from New Jersey, with which they compare closely, while the larger of the present specimens strongly resembles the *Menabites* (*Delawarella*) aff. *delawarensis* of Collignon (1948, pl. 30, fig. 2), if more involute. *M.* (*D.*) *campaniense* (de Grossouvre, 1894) is the only other species known from Europe (see below, p. 83). It differs from the present specimens in its much coarser ribbing (compare Pl. 12, figs. 3–6 and Pl. 11, figs. 3 and 4) with massive submarginal tubercles (3) and a depressed, polygonal whorl section, plus an effaced lateral tubercle (2) at a stage where it is still prominent in *delawarensis*.

Occurrence. Assise P^1 in northern Aquitaine, also known from the Lower Campanian of Delaware, New Jersey and the Gulf Coast region of the USA, Zululand and, perhaps, Madagascar.

Menabites (*Delawarella*) *campaniense* (de Grossouvre, 1894)

Plate 1, fig. 9; Plate 12, figs. 3–6; text-fig. 4B, D

1894 *Mortoniceras campaniense* de Grossouvre, p. 84 (*pars*), pl. 13, fig. 3 only, *non* 1 (= *M.* (*D.*) *delawarensis* (Morton, 1830)).
1901 *Mortoniceras delawarense* Morton; de Grossouvre, p. 901 (*pars*).
?1908*b* *Mortoniceras campaniense* de Grossouvre; de Grossouvre, p. 311.
1925 *Mortoniceras delawarense* Morton; Diener, p. 145 (*pars*).
1948 *Menabites* (*Delawarella*) *campaniensis* de Grossouvre; Collignon, p. 30 (87).
non 1963 *Delawarella campaniensis* (Grossouvre, 1894); Young, p. 113, pl. 64, figs. 2 and 6; pl. 67, fig. 2; text-figs. 24*a* and 25*a*.
non 1985 *Delawarella campaniensis* (de Grossouvre, 1894); Kennedy, pl. 3, figs. 3 and 4 (= *M.* (*D.*) *delawarensis* (Morton, 1830)).

Types. Lectotype, here designated, SP unregistered *ex* Arnaud Collection, from Assise P^1, St Hilaire-de-Montmoreau (Charente), the original of de Grossouvre 1894, pl. 13, fig. 3. The smaller figured paralectotype of de Grossouvre (1894, pl. 13, fig. 1) is a *Menabites* (*Delawarella*) *delawarensis* and is described above. Unfigured paralectotypes are two specimens, labelled 'Montmoreau P^2' (Pl. 1, fig. 9; Pl. 12, figs. 5 and 6) and 'Montmoreau P^{1-2}' (Pl. 11, figs. 7 and 8) the latter a *M.* (*Australiella*), all in the same collection. I have not seen the paralectotype from Contes-les-Pins (Alpes-Maritimes) mentioned by de Grossouvre (1894, p. 85), nor the unlocalized specimen said to be in the Musée de Périgueux.

Dimensions.

	D	*Wb*	*Wh*	*Wb*:*Wh*	*U*
Lectotype	— (—)	42·5(—)	36·8(—)	1·2	— (—)
Paralectotype	111·0(100)	48·5(43·7)	43·5(39·2)	1·1	39·5(35·6)

Description. Coiling evolute, umbilicus broad (35–38 % of diameter), moderately deep, with submarginal tubercle (3) housed in notch in umbilical wall of succeeding whorl. Whorl section depressed, polygonal in costal section, with greatest breadth at submarginal tubercle (3). Ornament consists of twenty strong, coarse, distant, blunt straight prorsiradiate primary ribs per whorl at a diameter of 100 mm. At the smallest diameter visible (44 mm approximately) there is a strong umbilical bulla (1), massive conical submarginal (3), and strong clavate external tubercle (5), the latter twice as numerous as the former. As size increases, a feeble lateral (2) tubercle develops, the submarginal (3) becomes greatly enlarged and gives rise to pairs of ribs that develop coarse conical marginal (4) and clavate external (5) tubercles; the venter between the latter is broad and flat, with a strong, blunt siphonal keel, entire in the lectotype (Pl. 12, fig. 3), but nodose in the paralectotype (Pl. 12, fig. 5).

Suture with massive bifid elements.

Discussion. M. (*D.*) *campaniense* is easily separated from *M.* (*D.*) *delawarensis*, the other species found in Aquitaine, by its massive whorl section and very coarse tuberculation. The specimen from the Dessau Member of the Austin Chalk, Texas figured by Young (1963, pl. 64, figs. 2 and 6; pl. 67, fig. 2; text-figs. 24*a* and 25*a*) bears little resemblance to the present specimen.

Occurrence. As for material. Also recorded from Contes-les-Pins (Alpes-Maritimes).

Subgenus AUSTRALIELLA Collignon, 1948, p. 64 (19)

[= *Austinites* Adkins, 1933, p. 407; *nom. nud.*]

Type species. By original designation: *Mortoniceras australe* Besairie, 1930, p. 638, pl. 64, fig. 2.

Menabites (*Australiella*) cf. *besairiei* (Collignon, 1938)

Plate 11, figs. 7 and 8; text-fig. 4F

compare:

1938 *Mortoniceras Besairiei* Collignon, p. 75 (25), pl. 4, fig. 2.
1948 *Australiella Besairiei* Coll. 1938; Collignon, p. 44 (101).
1970 *Australiella besairiei* Coll.; Collignon, p. 54, pl. 630, fig. 2327.

Material. SP unregistered *ex* Arnaud Collection, from Assise P^{1-2} at Montmoreau (Charente). The specimen is a paralectotype of *Menabites* (*Delawarella*) *campaniense* (de Grossouvre, 1894).

Dimensions.

D	*Wb*	*Wh*	*Wb*:*Wh*	*U*
120·0(100)	55·6(46·3)	39·5(32·9)	1·4	46·8(38·7)

Description. The specimen is a rather worn wholly septate fragment. Coiling very evolute, with a broad, quite deep umbilicus. Whorl section very depressed, with greatest breadth at submarginal (3) tubercle in costal section. Ornament consists of strong, blunt, coarse, distant straight prorsiradiate primary ribs, ten per half whorl. They arise at coarse umbilical bullae (1) and develop massive conical submarginal tubercles (3), which give rise to pairs of low, broad ribs; these may also occasionally branch, so that there are twice as many ventrolateral as lateral ribs, each bearing a strong external (5) tubercle on either side of a broad venter with a blunt, coarse, undulose siphonal keel.

Suture with massive bifid elements.

Discussion. Trituberculation to a large size plus external (5) tubercles twice as numerous as the others (1, 3) show this to be a *Menabites* (*Australiella*), a subgenus not previously recognized in Europe. The specimen is rather worn, and could belong to one of several of the twelve forms listed in Klinger and Kennedy (1980); indeed, I find it difficult to believe that the several co-occurring Madagascan 'species' actually merit specific separation. The closest resemblance, however, is to *M.* (*Australiella*) *besairiei* Collignon, 1938 (p. 75 (25), pl. 4, fig. 2), to which it is referred, although more depressed in whorl section, with no great confidence that this is specifically separate from *M.* (*A.*) *australis* (Besairie, 1930) and other species.

Occurrence. Assise P^2 at Montmoreau (Charente); Middle Campanian, Zone of *Delawarella subdelawarensis* and *Australiella australis* of Madagascar.

Family SPHENODISCIDAE Hyatt, 1900, p. 585
Subfamily SPHENODISCINAE Hyatt, 1900, p. 585
[= Libycoceratinae Zaborski, 1982, p. 306]

Genus SPHENODISCUS Meek, 1871, p. 298
[= *Austrosphenodiscus* Olsson, 1944, p. 266]

Type species. By original designation: *Ammonites lenticularis* Owen, 1852, p. 579 (*non* Phillips 1829, pl. 6, fig. 5) = *A. lobatus* Tuomey, 1856, p. 168.]

Diagnosis. Medium sized to large oxycones; probably dimorphic. Venter may round and whorl section broaden at maturity. Smooth or feebly ornamented with growth striae, radial folds on ribs or with feeble lateral and ventrolateral tubercles or ribs. Suture with E/L divided into three with two distinct adventitious lobes. Numerous auxiliary elements, saddles may be entire at end.

Discussion. I endorse Zaborski's comments on the difficulty of determining the relationships between the various species referred to the genus. It is quite impossible to state with any confidence the limits of species and whether or not the numerous widely scattered smooth forms are one or many species.

EXPLANATION OF PLATE 12

Figs. 1 and 2. *Hoplitoplacenticeras* (*Hoplitoplacenticeras*) *marroti* (Coquand, 1859). SP unregistered *ex* Arnaud Collection, labelled Assise P^3, Courgeac-Beauvais (Charente).

Figs. 3–6. *Menabites* (*Delawarella*) *campaniense* (de Grossouvre, 1894). 3 and 4, the lectotype, SP unregistered *ex* Arnaud Collection, from Assise P^1, St Hilaire-de-Montmoreau (Charente), original of de Grossouvre 1894, pl. 13, fig. 3; 5 and 6, a paralectotype, SP unregistered *ex* Arnaud Collection, from Assise P^2, Montmoreau (Charente).

All figs. × 1.

KENNEDY, *Hoplitoplacenticeras* (*Hoplitoplacenticeras*), *Menabites* (*Delawarella*)

Occurrence. Sphenodiscus ranges throughout the Maastrichtian, with records from northern Aquitaine and the Petites-Pyrénées in France, the Netherlands, Poland, Bulgaria, Israel, Arabia, Nigeria, Peru, Brazil, Madagascar, the US Gulf Coast and Western Interior, and Mexico. In Aquitaine it appears in what may be the uppermost Campanian at Maurens (Dordogne) and extends to low in the presumed Maastrichtian. Zaborski (1982) suggests that association with *Libycoceras* in the Middle East, Nigeria, and Peru indicates it to be Upper Campanian in these areas too. Correlation of the Campanian–Maastrichtian boundary in these regions is disputed, however.

Sphenodiscus ubaghsi de Grossouvre, 1894

Plate 14, figs. 1–5; text-fig. 5c

1894 *Sphenodiscus ubaghsi* A. de Grossouvre, n. sp., p. 141, pl. 9, figs. 4 and 6; text-fig. 60.
?1894 *Sphenodiscus rutoti* A. de Grossouvre, n. sp., p. 143, text-fig. 61.
1903 *Sphenodiscus ubaghsi* Grossouvre; Hyatt, p. 82.
?1903 *Sphenodiscus rutoti* Grossouvre; Hyatt, p. 83.
1908*a* *Sphenodiscus Ubaghsi*, de Grossouvre; de Grossouvre, p. 19, pl. 1, fig. 2.
?1908*a* *Sphenodiscus rutoti*, de Grossouvre; de Grossouvre, p. 19, pl. 1, fig. 3.
1925 *Sphenodiscus ubaghsi* Grossouvre; Diener, p. 233 (with additional synonymy).
?1925 *Sphenodiscus rutoti* Grossouvre; Diener, p. 232.
1984*b* *Sphenodiscus ubaghsi* de Grossouvre; Kennedy, p. 158.
?1984*b* *Sphenodiscus rutoti* de Grossouvre; Kennedy, p. 158.
1985 *Sphenodiscus ubaghsi* de Grossouvre; Kennedy, p. 106.
?1985 *Sphenodiscus rutoti* de Grossouvre; Kennedy, p. 106.
1986 *Sphenodiscus ubaghsi* de Grossouvre; Kennedy *in* Kennedy *et al.*, p. 1011, pl. 2, figs. 13–15; text-fig. 3*a*, *b*.

Types. Lectotype, by the subsequent designation of Kennedy *in* Kennedy *et al.* 1986, p. 1011, SP unregistered *ex* Arnaud Collection, the original of de Grossouvre 1894, pl. 9, fig. 4; text-fig. 60, from the 'Calcaire greseux à Thécidées. Assise R de M. Arnaud. Maurens (Dordogne)'. The specimen in the EMP collections from the 'environs de Maurens' has not been firmly identified (but see p. 88). Paralectotype is SP unregistered *ex* Arnaud Collection, the original of de Grossouvre 1894, pl. 9, fig. 6 from the 'Calcaire blanc (assise R de M. Arnaud), pierre de taille de Maurens (Dordogne)' (see discussion under occurrence below).

Description. The lectotype is an internal mould of three camerae; the paratype a worn internal mould of half a whorl of body chamber. Oxycone, smooth with tiny occluded umbilicus and acute venter. Whorl breadth to height ratio approximately 0·38 in holotype, 0·32 in paratype. Suture (text-fig. 5c) with E/L divided into three elements by two adventive lobes plus at least seven other saddles, two moderately incised, the remainder with broad entire terminal foliole.

Discussion. Sphenodiscus rutoti de Grossouvre, 1894 (p. 143, text-fig. 61) is from the same horizon and locality as *S. ubaghsi*. The holotype and only known specimen is lost, and de Grossouvre gave an illustration of a partial suture only. This differs from that of *S. ubaghsi* but, as Zaborski (1982) has shown, this is a variable feature in *Sphenodiscus*, and the two are probably variants of a single species. With such limited material it is quite impossible to assess the relationship of *ubaghsi* to the other smooth species of *Sphenodiscus*, as de Grossouvre (1908*a*, p. 19), Zaborski (1982), and others have

---

EXPLANATION OF PLATE 13

Fig. 1. *Sphenodiscus* sp. EMP 66/12 A.32.7 *ex* Marmot Collection, from Issac (Dordogne).
Figs. 2, 3, 6. *Pachydiscus (Pachydiscus) launayi* (de Grossouvre, 1894). SP unregistered *ex* Arnaud Collection, from Assise P^1, Maine-Bardon (Charente).
Fig. 4. *Trachyscaphites pulcherrimus* (Roemer, 1841). SP unregistered *ex* Arnaud Collection, from Assise Q, Neuvic (Dordogne).
Fig. 5. *Scaphites (Scaphites) hippocrepis* (DeKay 1828). SP unregistered *ex* Arnaud Collection, from Assise P^1–P^2, Livernant (Dordogne).
All figs. × 1.

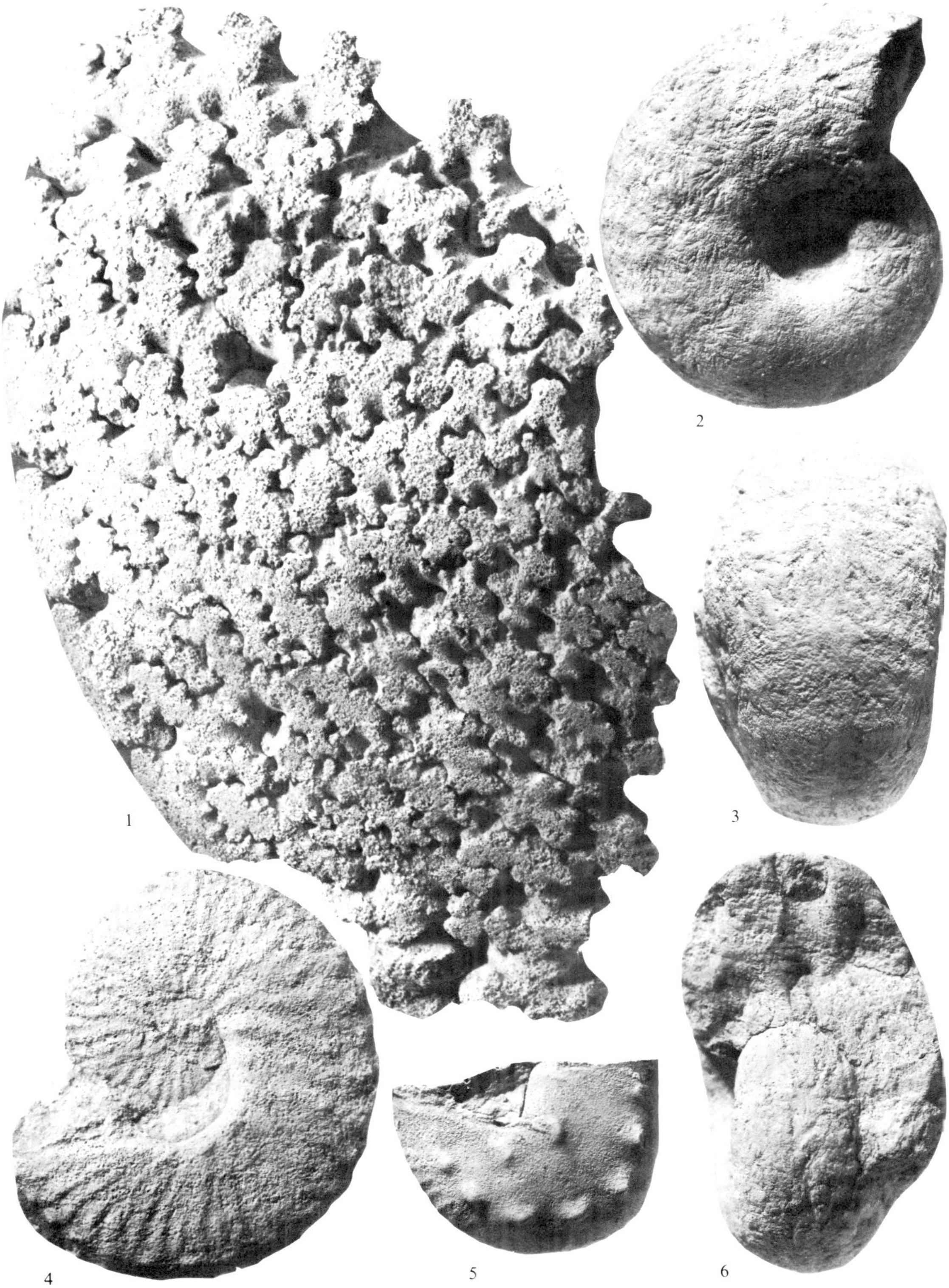

KENNEDY, *Sphenodiscus*, *Pachydiscus* (*Pachydiscus*), *Trachyscaphites*, *Scaphites* (*Scaphites*)

noted. The only other *Sphenodiscus* recorded from Europe is *S. binkhorsti* Böhm, 1898 (p. 197; see revision in Kennedy in press *b*) of the Upper Maastrichtian of the Netherlands, Poland, and Bulgaria. It is generally more compressed than *S. ubaghsi* and has a marked spiral angulation on the outer flank reflecting a change in whorl profile not seen in present form. Many *binkhorsti* show radial striae and faint ribs—or folds—that may strengthen into feeble outer lateral bullae or ribs in both juveniles and adults. The adventive lobes are much smaller than the lateral in *S. ubaghsi*, so that the adventive saddles are small with fewer incisions than in *S. binkhorsti* of the same size, while the auxiliary lobes are also less incised in *ubaghsi*.

Occurrence. The lectotype is described as coming from the 'Calcaires gréseux à Thecidées. Assise R de M. Arnaud' by de Grossouvre (1894, p. 142). Arnaud (1892) makes no mention of the specimen, suggesting that it was the one collected by de Grossouvre (1894, p. 141); but given de Grossouvre's precise indication that it was from the 'Calcaires gréseux à Thecidées' and the fact that Arnaud (1892, pp. 39 ff.) only records thecidaceans from his Bed *h* at Maurens, I suspect that this may be the source of the specimen. This bed corresponds to unit 3 of the detailed Maurens succession of Neumann *et al.* (1984) and is regarded as Lower Maastrichtian in microfossil terms. The surviving paralectotype from the 'Calcaire blanc (assise R de M. Arnaud), pierre de taille de Maurens' (de Grossouvre 1894, p. 142) is mentioned by Arnaud (1892, p. 142, l. 27) and precisely localized in his Bed *b*. Dr M. Neumann (Paris) has kindly compared the matrix of this specimen with the detailed succession described by Neumann *et al.* (1984) and tells me that the lithology is that of unit 7 or 8 of their sequence, and may thus be Upper Campanian (see discussion on p. 16). The only other specimens of *S. ubaghsi* are from the niveau de transition between the Marnes de Plagne and Calcaire nankin of Crabé and the Pas de Gazaille (Haute Garonne) (Kennedy *in* Kennedy *et al.* 1986). The niveau de transition yields presumed Lower Maastrichtian ammonites elsewhere, while the top of the Marnes de Plagne yields Upper Campanian *Bolivinoides draco-miliaris* 50 m from the summit succeeded by Lower Maastrichtian with *Globotruncana contusa*, *G. gansseri*, *G. patelliformis*, and *G. ventricosa*. The balance of evidence thus suggests *S. ubaghsi* is Lower Maastrichtian in microfossil terms and may first appear in the Upper Campanian.

Sphenodiscus sp.

Plate 13, fig. 1

Material. EMP 66/12 A.32.7 *ex* Marmot Collection, from Issac (Dordogne).

Discussion. This very large wholly septate fragment with a whorl height in excess of 90 mm is of great interest, given the rarity of *Sphenodiscus* in Aquitaine. Although rather battered the sutures show auxiliary lobes with more incisions than in *S. ubaghsi*, although this may reflect size difference.

Occurrence. Issac is 7 km east-south-east of Mussidan; the preservation of the specimen and locality suggest it to be from Assise R of Arnaud and of Upper Campanian or Lower Maastrichtian age.

Suborder ANCYLOCERATINA Wiedmann, 1966, p. 54
Superfamily TURRILITACEAE Gill, 1871, p. 3
Family NOSTOCERATIDAE Hyatt, 1894, p. 568

[= Jouaniceratidae Wright, 1952, p. 218; Bostrychoceratinae Spath, 1953, p. 16; Emperoceratinae Spath, 1953, p. 17; Hyphantoceratinae Spath, 1953, p. 16].

EXPLANATION OF PLATE 14

Figs. 1–5. *Sphenodiscus ubaghsi* de Grossouvre, 1894. Both SP unregistered *ex* Arnaud Collection, from Assise R of Arnaud, Maurens (Dordogne). 1–3, paralectotype; 4 and 5, lectotype.

Figs. 6 and 7. *Hoplitoplacenticeras* (*Hoplitoplacenticeras*) *rejaudryi* (de Grossouvre, 1894). SP unregistered *ex* Arnaud Collection, a paratype from Assise P^3, St Médard-de-Barbezieux (Charente).

Figs. 8 and 9. *Placenticeras bidorsatum* (Roemer, 1841). SP unregistered *ex* Arnaud Collection, from Assise P^2 inf., St Hilaire (Charente).

All figs. × 1.

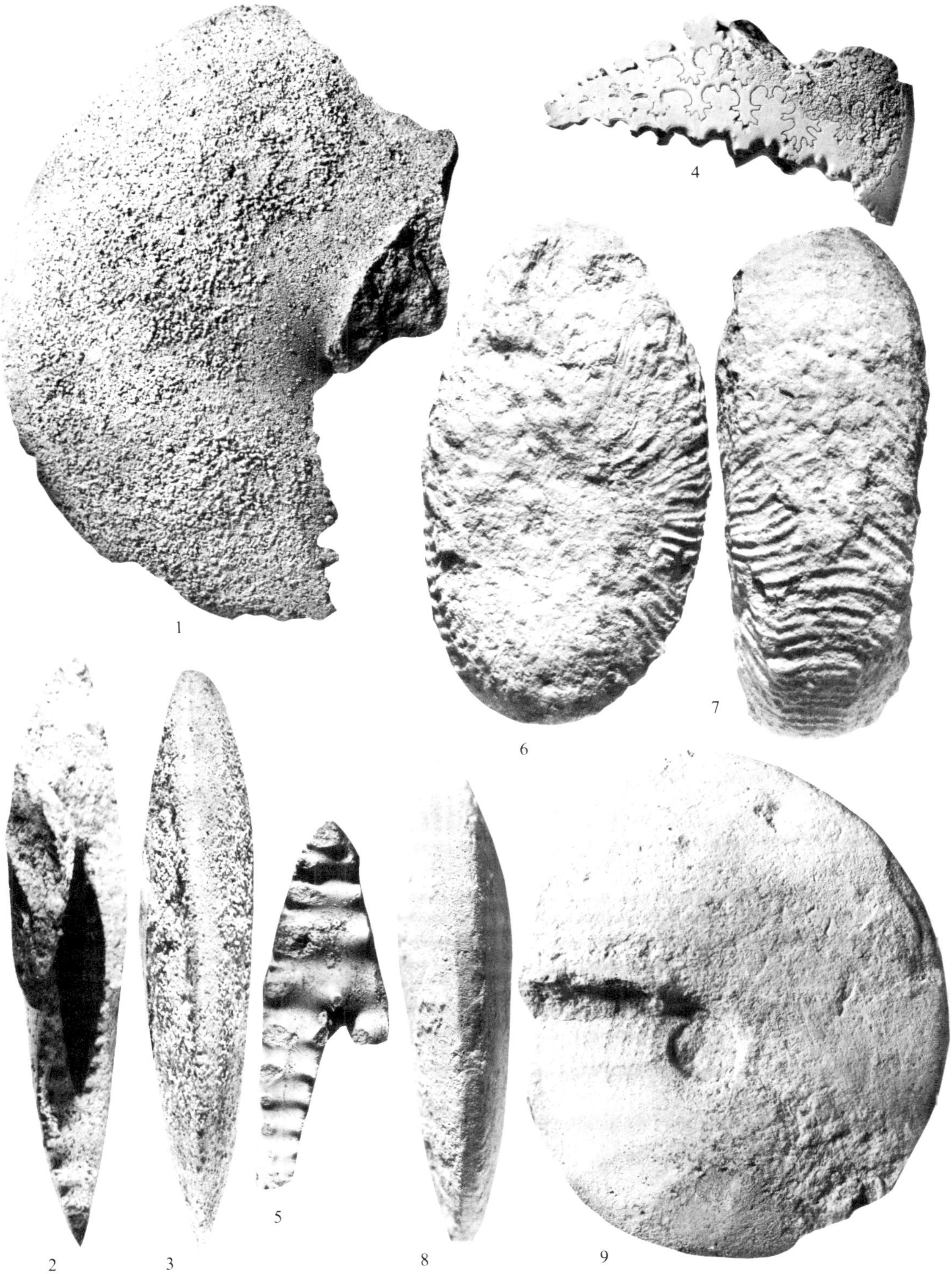

KENNEDY, *Sphenodiscus*, *Hoplitoplacenticeras* (*Hoplitoplacenticeras*), *Placenticeras*

Genus NOSTOCERAS Hyatt, 1894, p. 569

Type species. Nostoceras stantoni Hyatt, 1894, p. 569, by original designation.

Subgenus NOSTOCERAS Hyatt, 1894, p. 569

Nostoceras (*Nostoceras*) *hyatti* Stephenson, 1941

Plate 20, figs. 7–9

1941 *Nostoceras hyatti* Stephenson, p. 410, pl. 81, figs. 9–12.
1974 *Nostoceras hyatti* Stephenson; Cobban, p. 10, pl. 5, figs. 1–21; pl. 6, figs. 1–12; pl. 7, figs. 1–10; pl. 8, figs. 1–30; text-fig. 8 (with full synonymy).

Holotype. USNM 77258, the original of Stephenson 1941, pl. 81, fig. 9, from the Nacatoch Sand at Corsicana, Navarro County, Texas.

Material. SP unregistered *ex* Arnaud Collection, from Assise R^2, Maurens (Dordogne), cited by Arnaud as *Turrilites polyplocus* from his Bed *h*, correlated with Bed 3 of Neumann *et al.* (1984) and confirmed independantly from microfauna in the matrix (M. Neumann *in litt.*).

Description. Specimen is a wholly sepatate composite mould of a low helix with a maximum diameter of 28·5 mm. Whorl section depressed, oval with whorls loosely in contact; thirty-five ribs per whorl arise on the inner whorl face, pass across the upper whorl face and increase by branching or intercalation to give a total of fifty-three to fifty-five ribs per whorl. Ribs narrow and delicate, crowded, sweeping back and concave on the upper, outer whorl face, passing straight and prorsiradiate across the middle and lower outer whorl face and crossing the lower face more or less straight and radial. Ribs link in pairs at small upper outer lateral tubercles with a thickened rib leading to a lower outer lateral tubercle, with non-tuberculate intercalatories that may link to the lower tubercle. Strong distant constrictions, three per half whorl, are prominent. Ribbing is disturbed and irregular on the last half whorl.

Sutures not seen.

Discussion. Nostoceras hyatti is discussed in detail by Cobban (1974), who has kindly examined a cast of this specimen, confirming it as a low-spired member of this typically North American species. The reader is referred to his work for a full discussion of the limits and variation of *N. hyatti*.

Occurrence. The Aquitaine specimen is Lower Maastrichtian in microfossil terms, as discussed on p. 16. Other occurrences of *N. hyatti* are difficult to place with certainty in Upper Campanian versus Lower Maastrichtian; the geographic range extends from the eastern Seaboard and Gulf Coast in the United States to Angola, possible Israel and Madagascar.

Nostoceras (*Nostoceras*) *archiacianum* (d'Orbigny, 1842)

Text-fig. 31B–E, H, I

1842 *Turrilites archiacianus* d'Orbigny, p. 607, pl. 148, figs. 5 and 6.
1850 *Turrilites archiacianus* d'Orbigny; d'Orbigny, p. 216.
?1913 *Heteroceras polyplocum* Roemer var. *Schlönbachi* Favre sp.; Nowak, p. 385, pl. 41, fig. 14; pl. 44, fig. 42.
1921 *Didymoceras archiacianum* d'Orbigny sp.; Spath, p. 249.
?1934 *Bostrychoceras polyplocum* Roemer; Zhelev, p. 196, pl. 4, fig. 5 only.
?1951 *Bostrychoceras polyplocum* var. *schloenbachi* (Favre); Mikhailov, p. 51, pl. 2, figs. 13 and 14; pl. 3, figs. 16 and 19; ?*non* pl. 4, figs. 20–22.
?1951 *Bostrychoceras polyplocum* var. *renngarteni* Mikhailov, p. 54, pl. 3, figs. 17 and 18.
1958 *Turrilites* (*Bostrychoceras*) *archiacianus* d'Orbigny; Sornay, p. 1, figs. 1 and 2.
?1959 *Bostrychoceras schloenbachi* (Favre); Naidin and Shimanskij, p. 180, pl. 4, figs. 1–5.
1962 *C.* (*Cirroceras*) *schloenbachi schloenbachi* (Favre); Wiedmann, p. 204, pl. 11, fig. 2.
?1964 *Bostrychoceras polyplocum schloenbachi* Favre; Tzankov, p. 151, pl. 3, figs. 2 and 3.
?1974 *Bostrychoceras schloenbachi* (Favre, 1869); Naidin, p. 166 (*pars*), pl. 55, figs. 2–7, *non* 8.
?1980 *Nostoceras* (?) *schloenbachi* (Favre, 1869); Błaszkiewicz, p. 27, pl. 10, figs. 6, 7, 10.
1984 *Nostoceras* (*Nostoceras*) sp.; Kennedy and Summesberger, p. 163, pl. 9, figs. 1, 2, 3, 9.

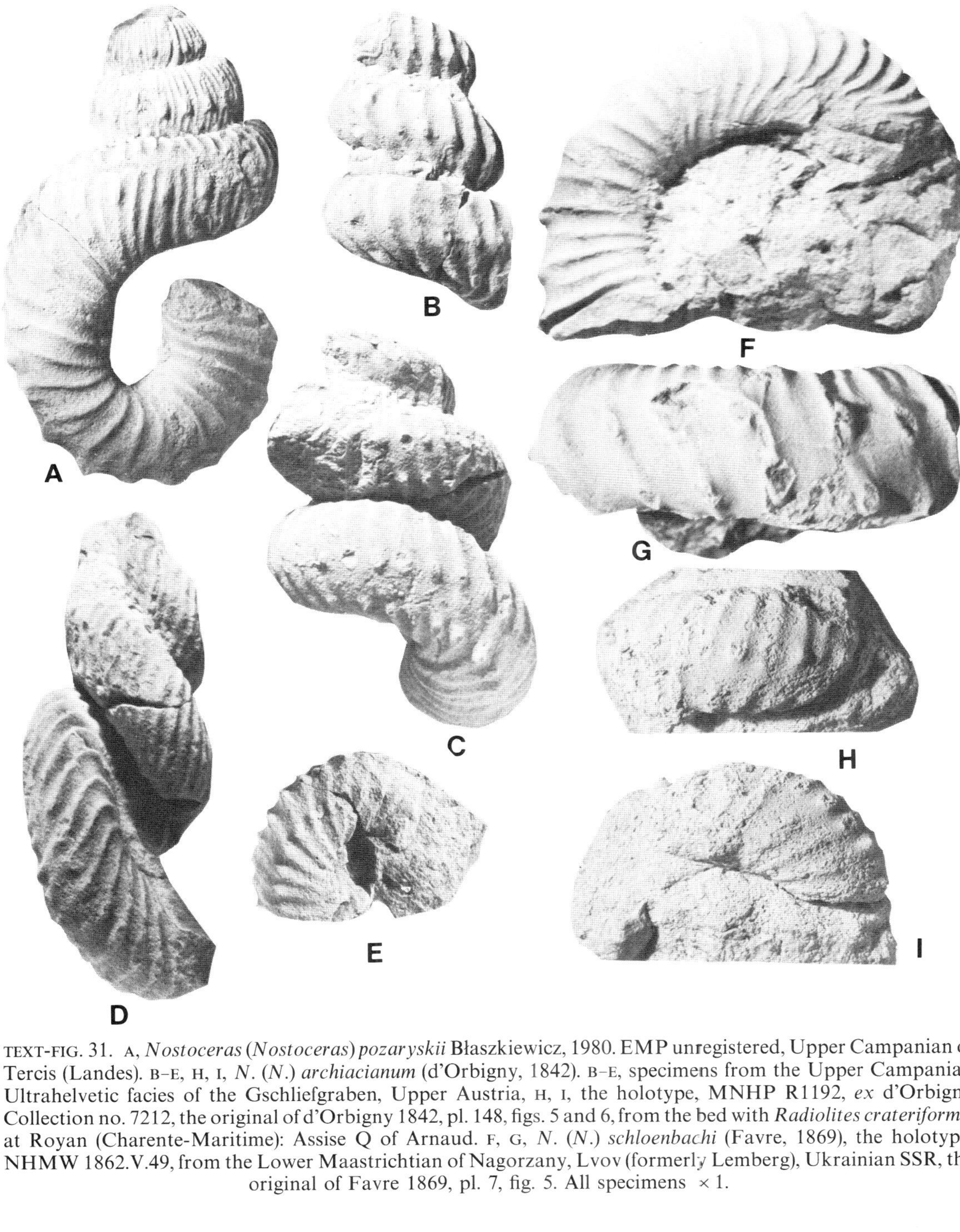

TEXT-FIG. 31. A, *Nostoceras* (*Nostoceras*) *pozaryskii* Błaszkiewicz, 1980. EMP unregistered, Upper Campanian of Tercis (Landes). B–E, H, I, *N.* (*N.*) *archiacianum* (d'Orbigny, 1842). B–E, specimens from the Upper Campanian Ultrahelvetic facies of the Gschliefgraben, Upper Austria, H, I, the holotype, MNHP R1192, *ex* d'Orbigny Collection no. 7212, the original of d'Orbigny 1842, pl. 148, figs. 5 and 6, from the bed with *Radiolites crateriformis* at Royan (Charente-Maritime): Assise Q of Arnaud. F, G, *N.* (*N.*) *schloenbachi* (Favre, 1869), the holotype, NHMW 1862.V.49, from the Lower Maastrichtian of Nagorzany, Lvov (formerly Lemberg), Ukrainian SSR, the original of Favre 1869, pl. 7, fig. 5. All specimens × 1.

Type. Holotype, by monotypy, MNHP R1192 *ex* d'Orbigny Collection no. 7212, from the bed with *Radiolites crateriformis* at Royan (Charente-Maritime): Assise Q of Arnaud.

Description. The holotype is an internal mould of part of a helically coiled section of body-chamber with a maximum preserved whorl height of 18·5 mm. Strong, distant prorsiradiate primary ribs are widely spaced. They sweep forwards over the upper, outer flank, and strengthen into bullate tubercles, present on all ribs low on the flank. There is some variation in tubercle strength, while the ribs weaken on the lower part of the flank before strengthening into a smaller tubercle at the junction of the outer and the lower whorl face. The ribs sweep forwards and are feebly convex on the lower whorl face. There is some irregularity in the ribs, a few declining on the outer whorl face.

Discussion. D'Orbigny's figure is accurate, if upside-down. This has been a neglected species, but examination of the holotype shows it to correspond to the more complete specimens described as *Nostoceras* (*Nostoceras*) sp. from the *polyplocum* Zone fauna of the Ultrahelvetic of the Gschliefgraben (Upper Austria) (Kennedy and Summesberger 1984), shown in text-fig. 31B–E. The simple ribbing and tuberculation of the holotype and Austrian material suggest that at least some of the Upper Campanian material from Europe and the USSR referred to as *Bostrychoceras polyplocum renngarteni* Mikhailov, 1951 (p. 54, pl. 3, figs. 17 and 18) and *N.* (*N.*) *schloenbachi* (Favre, 1869) (e.g. Mikhailov 1951, pl. 2, figs. 13 and 14; Wiedmann 1962, pl. 11, fig. 2) belong here. With limited material it is not possible to determine whether specimens with intercalated non-tuberculate ribs or with ribs linking in pairs to tubercles are variants of the present species or not (e.g. the *N.* (?) *schloenbachi* of Błaszkiewicz, 1980, p. 27, pl. 10, figs. 6, 7, 10 (with synonymy)). Certainly, as Błaszkiewicz (1980, p. 27) and Kennedy and Summesberger (1984, p. 164) note these Campanian forms are not true *schloenbachi*, a Lower Maastrichtian species (text-fig. 37F, G). Also distinct from the present material and from *schloenbachi* are loose helices attributed to *schloenbachi* by Wiedmann (1962) and Klinger (1976) (see below). This material apart, the other important western European *N.* (*Nostoceras*) is *pozaryskii* Błaszkiewicz, 1980 (p. 26, pl. 10, figs. 1–5, 8, 9, 11–15) index of the top zone of the Campanian in Poland. The types are poor, but a fine Tercis (Landes) specimen shown in text-fig. 31A clarifies differences from *N.* (*N.*) *archiacianum*. The end of the phragmocone and early body-chamber thus show fine ribs looping between the tubercles with some non-tuberculate intercalatories, the later part of the body-chamber is distantly ribbed, the ribs bearing long clavi joining ribs in pairs and zig-zagging between adjacent rows of tubercles (Błaszkiewicz 1980, pl. 10, fig. 5).

Occurrence. As for type. The species also occurs in the Upper Campanian *polyplocum* Zone fauna of the Ultrahelvetic, Gschliefgraben, Upper Austria and, possibly, more widely in northern Europe and the USSR.

Subgenus BOSTRYCHOCERAS Hyatt, 1900, p. 588

Type species. Turrilites polyplocus Roemer, 1841, p. 92, pl. 14, figs. 1 and 2, by original designation.

Nostoceras (*Bostrychoceras*) *polyplocum* (Roemer, 1841)

Plate 6, fig. 1; Plate 15, figs. 1–3, 5–8

1841 *Turrilites polyplocus* Roemer, p. 92, pl. 14, fig. 1 only, *non* 2 (= *Eubostrychoceras saxonicum* (Schlüter, 1876)).

EXPLANATION OF PLATE 15

Figs. 1–3 and 5–8. *Nostoceras* (*Bostrychoceras*) *polyplocum* (Roemer, 1841). 1, 3, 7, SP unregistered *ex* Arnaud Collection, from Assise P^3, Caillaud (Charente-Maritime); 2, 5, 6, P. Moreau Collection (Poitiers), from Assise P^3, Zone Industrielle, Saint Georges, Royan (Charente-Maritime); 8, Université de Poitiers Collections, from Assise Q, Archiac (Charente-Maritime).

Fig. 4. *Trachyscaphites pulcherrimus* (Roemer, 1841). SP unregistered *ex* Arnaud Collection, from Assise P^3, St Médard [-de-Barbezieux] (Charente), identified as *Bostrychoceras polyplocum* by Arnaud.

All figs. × 1.

KENNEDY, *Nostoceras* (*Bostrychoceras*), *Trachyscaphites*

non 1843 *Turrilites polyplocus* Röm; Geinitz, p. 67, pl. 3, fig. 1 (= *Eubostrychoceras saxonicum* (Schlüter, 1876))
non 1846 *Turrilites polyplocus* Röm; Geinitz, p. 12, fig. 3 (= *Eubostrychoceras saxonicum* (Schlüter, 1872*a*)).
1850 *Heteroceras polyplocus* d'Orbigny, p. 216.
non 1850 *Turrilites germaniae* d'Orbigny, p. 216.
1872*a* *Heteroceras polyplocum* A. Roemer sp.; Schlüter, p. 112, pl. 33, figs. 3–8; pl. 34, figs. 1–5; pl. 35, figs. 1–8.
1889 *Heteroceras polyplocum* A. Roemer; Griepenkerl, p. 407, pl. 12, fig. 1.
1902 *Heteroceras polyplocum* A. Roemer sp.; Wollemann, p. 97.
1904 *Heteroceras polyplocum* Roemer; Douvillé, p. 256, pl. 35, fig. 2.
non 1907 *Turrilites* (*Bostrychoceras*) *polyplocus* Roemer; Boule *et al*., p. 41 (61), pl. 7 (14), figs. 1 and 2 (= *Eubostrychoceras indopacificum* Matsumoto, 1967; *fide* Matsumoto).
1907 *Bostrychoceras polyplocum* Roemer; Pervinquière, p. 102.
1911 *Bostrychoceras polyplocum* (Roemer); Lopuski, p. 183.
non 1913 *Heteroceras polyplocum* Roemer sp.; Nowak, p. 385, pl. 40, fig. 6; pl. 45, fig. 48 (? = *Bostrychoceras* (*Mobergoceras*) sp.).
1925 *Bostrychoceras polyplocum* A. Roemer; Diener, p. 87 (*pars*) (with additional synonymy).
1928 *Bostrychoceras* n. sp. aff. *polyplocum* (Roemer); Adkins, p. 214, pl. 37, figs. 1 and 3.
1934 *Bostrychoceras polyplocum* Zhelev, p. 194, pl. 40, fig. 6 only.
1936 *Turrilites undulatus* Sow.; Nalivajko, p. 36, pl. 15, fig. 43.
1938 *Bostrychoceras polyplocum* Roemer; Roman, p. 445, pl. 45, fig. 425.
1951 *Bostrychoceras polyplocum* (Roemer); Mikhailov, p. 50, pl. 4, figs. 25–27; pl. 5, fig. 31.
1957 *Bostrychoceras polyplocum* (Roemer); Wright, p. L224, fig. 250, 4.
1959 *Bostrychoceras polyplocum* (Roemer); Naidin and Shimanskij, p. 180, pl. 4, fig. 6; pl. 5, figs. 1 and 2.
1960 *Bostrychoceras* cf. *polyplocum* (Roem); Wiedmann, p. 716.
1962 *C.* (*Cirroceras*) *polyplocum* (A. Roemer); Wiedmann, p. 198.
1962 *C.* (*Cirroceras*) *depressum* Wiedmann, p. 199.
1962 *C.* (*Cirroceras*) *polyplocum polyplocum* (A. Roemer); Wiedmann, p. 200, pl. 11, fig. 4.
non 1962 *C.* (*Cirroceras*) *polyplocum zumayense* Wiedmann, p. 200, pl. 9, fig. 5 (undeterminable).
1962 *C.* (*Cirroceras*) *indicum* (Stol.); Wiedmann, p. 200 (*pars*).
?1962 *C.* (*Cirroceras*) *indicum saxonicum* (Schlüt); Wiedmann, p. 203.
non 1962 *C.* (*Cirroceras*) *indicum elongatum* (Whiteaves); Wiedmann, p. 203, pl. 9, fig. 4.
1962 *C.* (*Cirroceras*) *schloenbachi densecostatum* Wiedmann, p. 204.
1963 *Bostrychoceras secoense* Young, p. 42, pl. 3, figs. 1–5; pl. 4, figs. 4 and 8; text-fig. 7s.
1974 *Bostrychoceras polyplocum* (Roemer, 1841); Naidin, p. 165, pl. 54, figs. 1–3, pl. 55, fig. 1.
1975 *Bostrychoceras* (*Bostrychoceras*) *polyplocum* (F. A. Roemer, 1841); Schmid and Ernst, p. 339, pl. 4, figs. 2 and 3.
1976 *Bostrychoceras polyplocum* (Roemer); Atabekian and Khakhimov, p. 54, pl. 6, figs. 1–4; pl. 7, figs. 1 and 2.
1980 *Bostrychoceras polyplocum polyplocum* (Roemer, 1841); Błaszkiewicz, p. 20, pl. 1, figs. 1–9; pl. 11, figs. 2, 3, 5, 6.
1980 *Bostrychoceras polyplocum schlueteri*; Błaszkiewicz, p. 20, pl. 2, figs. 1, 4, 9–11.
1980 *Bostrychoceras unituberculatum* Błaszkiewicz, p. 21, pl. 3, figs. 1–8; pl. 4, figs. 3–6.
?*non* 1980 *Didymoceras densecostatum* (Wiedmann, 1962); Błaszkiewicz, p. 25, pl. 7, figs. 2–4.
1980 *Didymoceras* cf. *secoense* (Young, 1963); Błaszkiewicz, p. 24, pl. 5, figs. 4 and 6; pl. 7, figs. 16 and 19.
?1980 *Didymoceras* sp. Błaszkiewicz, p. 22, pl. 4, figs. 1 and 2; pl. 5, fig. 7.
?1980 *Didymoceras* cf. *beecheri* Hyatt; Błaszkiewicz, p. 22, pl. 5, figs. 3 and 5.
?1980 *Didymoceras varium* Błaszkiewicz, p. 22, pl. 6, figs. 1–7; pl. 7, figs. 21 and 22.
?1980 *Didymoceras postremum* Błaszkiewicz, p. 25, pl. 8, figs. 1–6; pl. 9, figs. 1–6.
1982 *Bostrychoceras polyplocum* (Roemer); Martinez, p. 164, pl. 28, figs. 1–3.
1982 *Bostrychoceras polyplocum* (Roemer, 1841); Tzankov, p. 18, pl. 3, fig. 5.
1984 *Bostrychoceras polyplocum* Roemer; Bilotte, pl. 37, fig. 4.
1984 *Nostoceras* (*Bostrychoceras*) *polyplocum* (Roemer, 1841); Kennedy and Summesberger, p. 164, pl. 9, figs. 4, 8, 12–14.
1985 *Nostoceras* (*Bostrychoceras*) *polyplocum* (Roemer, 1841); Kennedy, pl. 3, figs. 8 and 11.

Types. Roemer mentions specimens from Dülmen, Lemförde, and Weinböhlen, figures one from an unspecified locality (pl. 14, fig. 1) and a second from 'Pläner bei Weinböhlen' (pl. 14, fig. 2). In the synonomy he cites 'Geinitz pl. 13, fig. 2', the original of which is thus also a syntype. Wiedmann (1962, p. 198) states that Schlüter (1872*a*) designated the original of Roemer's pl. 14, fig. 1 lectotype, but this is not the case; Błaszkiewicz (1980, p. 20) attributes lectotype designation to Wiedmann (1962, p. 198), but this is also not the case. Roemer's types are lost; the only specimen assignable to *polyplocum* in the collections at Wroclaw (letter from Dr. J. Gorczyca-Skala, 27.11.1984) is from Haldem and is thus not part of the type series. The paratype figured by Geinitz, is, according to Wiedmann (1962, p. 198) referable to *Turrilites saxonicus* Schlüter, 1876 (p. 135), who designated the lost original of Roemer 1841, pl. 14, fig. 2, the lectotype of *saxonicus*. In the interests of nomenclatural stability I here designate the original of Roemer 1841, pl. 14, fig. 1, lectotype of *polyplocum*. It is from an unspecified locality in north Germany. No less than five specific names have been applied to north German *polyplocum* and neotype designation is highly desirable to stabilize the nomenclature.

Material. P. Moreau Collection (Poitiers): 6 specimens from Assise P^3, Zone Industrielle, Saint-Georges, Royan (Charente-Maritime); Université de Poitiers unregistered, 1 specimen from Assise Q, Archaic (Charente-Maritime); SP unregistered *ex* Arnaud Collection: 1 specimen from Assise P^3, Courgeac (Charente); 1 specimen from assise P^3 inf., Lande de Heugas; 3 specimens from Assise P^3, Caillau Supérieur (Charente-Maritime); 1 specimen from Assise Q, Rioux Martin (Charente).

Description. The Aquitaine material is fragmentary. Coiling is dextral or sinistral, whorl section circular; whorls loosely in contact. Ornament consists of numerous dense, crowded, rounded, flexuous prorsiradiate ribs. These are concave on the upper outer flank but flex and are concave on the lower outer flank. Tuberculation is very variably developed; some specimens show an occasional upper lateral tubercle only (Pl. 6, fig. 1), where ribs branch, or may lack them altogether (Pl. 15, fig. 2). Others have well-developed upper lateral tubercles plus a second row at the juncture of outer and lower whorl faces. Ribs link in pairs at, and loop between tubercles, with additional non-tuberculate ribs intercalated between (Pl. 15, fig. 5). The largest specimen is part of the coiled section of the body-chamber. It bears a single row of tubercles only, at the juncture of the outer and lower whorl faces (Pl. 15, fig. 8), whence ribs link in pairs, while there are also branching and intercalated ribs on the outer face.
None of the specimens show the sutures.

Discussion. The Aquitaine material has not been illustrated or described before, although a *polyplocum* Zone has been in use since the early part of the century. It is inadequate as a basis for discussing this variable species, which has been subject to varying and conflicting interpretations by recent revisers, notably Wiedmann (1962) and Błaszkiewicz (1980). I recently studied the large collections of more than 100 specimens of this species from Haldem (Westphalia) in the Münster Collection (Museum für Naturkünde, Berlin) and the Schlüter Collection (Bonn) in order to assess variation and development in the species. These show *Bostrychoceras polyplocum* to be markedly dimorphic, micro- and macroconch body-chambers occurring in a near 1:1 ratio. The material is crushed, but the mouth border of microconchs ranged from 38·5 to 44·0 mm in height; in macroconchs the corresponding figures are 54–65 (text-fig. 32). As a result, fragments of juveniles at the same size may represent different ontogenetic stages. Coiling is both dextral and sinistral. All the small specimens seen, down to a whorl height of 10 mm are loosely coiled (text-fig. 34A–F). Depending on the degree of crushing, the apical angle varies widely. Whorl section appears to have been circular. Ornament consists of dense ribs, concave on the upper outer whorl face, sweeping forwards and crossing the outer whorl face in a sinusoidal curve, convex on the mid to lower flank and sweeping back on the underside of the whorl. Because of crushing it is difficult to determine rib density: a figure of eighty to ninety per whorl is, perhaps, typical. These early whorls invariably bear tubercles. In general, two rows are present, one low on the outer whorl face, the other at the juncture of outer and lower whorl faces. Ribs link together in pairs at the upper tubercle and connect to the lower in 'button and loop' fashion; a single non-tuberculate rib, sometimes two, are inserted between the tuberculate, looped pairs (text-fig. 34B). Widely spaced constrictions, two to three per whorl, are preceded by a flared rib.

This pattern of ornament typifies all specimens seen (30) to a diameter of 20 mm. Beyond this, coiling tightens, and the whorls are in contact. In some specimens the bituberculate condition extends to only 20–35 mm whorl height, whereafter all tuberculation is abruptly lost, leaving a shell ornamented by branching ribs only; this ornament extending to the final recurved hook (text-fig. 32).

TEXT-FIG. 32. A–B, *Nostoceras* (*Bostrychoceras*) *polyplocum* (Roemer, 1841). A, adult microconch, GPIB 72*a*, the original of Schlüter 1872*a*, pl. 33, fig. 5; B, an adult macroconch, MNB unregistered *ex* Münster Collection; both from the Upper Campanian of Haldem, Westphalia, both ×0·8.

In other specimens the tubercles persist in a rather irregularly spaced fashion, with some specimens perfectly bituberculate throughout (text-fig. 33A), some with occasional lower or occasional upper tubercles (text-fig. 35A), some with upper row alone and perfectly unituberculate (text-fig. 35B), some with a few lower tubercles added. Constrictions persist, or not, to a whorl height of 45 mm.

The wide variation in the stage at which bituberculation is lost is clearly a reflection of the size dimorphism. Some middle growth stage specimens have both bi- and uni-, or bi- and non-tuberculate portions (text-fig. 35A).

At maturity the last whorl uncoils into an irregular U-shaped hook (text-fig. 32). All specimens show a renascence of a pair of distant irregular coarse tubercles linking one to three ribs, while the adult mouth border is preceded by a markedly flared rib bearing a pair of tubercles, succeeded by a short contracted tubular aperture.

The French material from P^3, although limited, clearly shows the same spread of variation. The two specimens from Q, in contrast, lack tubercles, but the sample is too small to justify according significance to this.

On the basis of study of the Haldem material I can see no lines of specific or subspecific division. It must be admitted that the Haldem succession is around 100 m thick, and the ammonite fauna shows this to span much of the Upper Campanian. Following the work of Schmid and Ernst (1975), *polyplocum* has only a limited range in north Germany, and I see no reason to believe other than the material coming from only the higher part of the Haldem sequence, as shown by Arnold (1968).

With these observations in mind, the missing lectotype (Roemer 1841, pl. 14, fig. 1) is seen to be an individual lacking tubercles in middle growth. The apparent tubercles are irregularly placed, and are no more than conellae like those developed on the large Haldem specimen shown in text-fig. 32. *T. germaniae* d'Orbigny, 1850 (p. 216), regarded as a synonym by Wiedmann (1962, p. 198) is described as having more than two rows of tubercles. No specimens survive in the d'Orbigny Collection and I regarded it as a *nomen dubium*; if accurately described it cannot be a *polyplocum*, although from Haldem. Of specimens illustrated by Schlüter (1872*a*) from Haldem, I refigure the originals of pl. 33, figs. 5–8 here as text-figs. 32A and 34B–F. All are *polyplocum*. Wiedmann (1962, p. 199) regarded the missing originals of pl. 33, figs. 3 and 4 as belonging to the Indian species *Heteroceras indicum* Stoliczka, 1866 (p. 184, pl. 86, figs. 1 and 2), subspecies *saxonicum* Schlüter, 1876 (p. 135).

The syntypes of *indicum* were refigured by Matsumoto (1967, pl. 18, figs. 2 and 3) and referred to *Eubostrychoceras* Matsumoto, 1967 with a query: a fine Madagascan specimen is shown in text-fig. 36. They bear only superficial similarity to the Haldem specimens, and I take Schluter's pl. 33, figs. 3 and 4 to be absolutely typical nontuberculate *polyplocum*. The original of Schlüter's pl. 34, fig. 1 is shown in text-fig. 33A. It was referred to *polyplocum* by Wiedmann (1962, p. 198), and is a variant with well-developed bituberculation throughout. The original of pl. 34, figs. 2 and 3 is missing, and is the holotype of *Cirroceras depressum* Wiedmann, 1962 (footnote, p. 199). From Haldem, it seems to me to be no more than a *polyplocum* that has been buried in a vertical position and crushed at 90° to the coiling axis whereas most specimens are crushed normal to this axis (cf. Schmid and Ernst 1975, p. 340). A comparable individual is shown in text-figs. 34G, H and 35D. Lacking tubercles, it closely resembles Schlüter's pl. 34, figs. 4 and 5. Wiedmann describes *depressum* as having 120 ribs per whorl; the figure of the holotype shows just over 100. The giant *Didymoceras* (*Didymoceras*) *depressum sanctaeluciense* Klinger, 1976 (p. 65, pls. 25–29; pl. 29, figs. 1–3; pl. 30, figs. 1 and 2; text-fig. 8*f*, *g*, *h*) is a Maastrichtian form with a truly wide apical angle, unrelated to the Campanian *polyplocum*. The original of Schlüter's pl. 35, figs. 1–4, referred to by him as *Heterecoceras polyplocum*? was made the holotype of *C.* (*Cirroceras*) *schloenbachi densecostatum* Wiedmann, 1962; this specimen (GPIB unregistered) is shown in text-fig. 33C–E. The original of Schlüter's pl. 35, figs. 5–7 was made a paratype and survives as GPIB unregistered. The holotype appears to be a fragment of *polyplocum* crushed along the axis of coiling. It shows the same ornament as specimens regarded by previous authors as typical *polyplocum* and the coarse flares match closely with the original of Schlüter 1872, pl. 34, fig. 1, shown in text-fig. 33A. I equally see no reason for regarding the paratype as other than a juvenile *polyplocum*. True *Helicoceras schloenbachi* Favre, 1869 (p. 30, pl. 7, fig. 5) has been generally

TEXT-FIG. 33. A–E, *Nostoceras* (*Bostrychoceras*) *polyplocum* (Roemer, 1841). A, GPIB unregistered, the original of Schlüter 1872*a*, pl. 34, fig. 1; B, cast of the specimen figured by Young (1963, pl. 4, fig. 8) as *Bostrychoceras secoense* Young, 1963, Upper Campanian of Battle Ground Creek, Travis County, Texas. University of Texas Collections, no. 1613; C–E, holotype of *Ciroceras* (*Cirroceras*) *schloenbachi densecostatum* Wiedmann, 1962, the original of Schlüter 1872*a*, pl. 35, figs. 1–4, GPIB unregistered; A, C–E, Upper Campanian of Haldem, Westphalia, German Federal Republic. All specimens ×1.

TEXT-FIG. 34. A–H, *Nostoceras* (*Bostrychoceras*) *polyplocum* (Roemer, 1841). A, MNB unregistered; B, C, F, GPIB 72*b*, the original of Schlüter 1872*b*, pl. 33, fig. 6; D, E, GPIB 72*c*, the original of Schlüter 1872*b*, pl. 33, figs. 7 and 8; G, H, MNB unregistered. All specimens from the Upper Campanian of Haldem, Westphalia, German Federal Republic, ×1.

TEXT-FIG. 35. A–D, *Nostoceras* (*Bostrychoceras*) *polyplocum* (Roemer, 1841). All are unregistered MNB specimens from the Upper Campanian of Haldem, Westphalia, German Federal Republic, ×1.

TEXT-FIG. 36. *Nostoceras* (*Eubostrychoceras*) *indopacificum* Matsumoto, 1967. The original of Boule *et al.* 1907, pl. 7 (14), figs. 1 and 2, an adult macroconch from the 'Sénonien' of the Montagne des Français, Madagascar; MNHP Collections, ×0·75.

misinterpreted. The holotype (text-fig. 31F, G) is part of the body-chamber of a *Nostoceras* (*Nostoceras*) from the Lower Maastrichtian of Lvov (formerly Lemberg) in the Ukranian SSR (formerly Galicia). The original of Schlüter's pl. 34, fig. 1 has been regarded as identical with *Bostrychoceras secoense* Young, 1963 (p. 42, pl. 3, figs. 1–5; pl. 4, figs. 4 and 8; text-fig. 7*s*) by Young, a view with which I concur, placing *secoense* in synonymy as a result (compare text-fig. 33A and B). Błaszkiewicz (1980, p. 25, pl. 7, figs. 2–4) figures what may be a different form as *D. densecostatum*; it is from the upper Upper Campanian of the Vistula Valley, Poland. The original of Schlüter's pl. 35, fig. 8 is missing. It shows loose coiling with the whorls widely separated. I was unable to recognize the original in the Bonn Collections but figure a similarly coiled but persistently tuberculate individual as text-fig. 34A. Both I consider no more than coiling variants of *polyplocum*.

Madagascan *polyplocum* (Boule *et al.*, 1907, p. 41 (61), pl. 7 (14), figs. 1 and 2) are *Eubostrychoceras*; referred to *saxonicum* of Schlüter by Wiedmann (1962, p. 201), Matsumoto places them in his *indopacificum* (1967, p. 333). This subgenus is markedly dimorphic (the holotype of *indopacificum* is a microconch: Matsumoto 1967, pl. 18, fig. 1; the larger of Boule, Lemoine, and Thevénin's examples is a macroconch: text-fig. 36); the species differs from *polyplocum* in its total lack of tubercles, lack of a loose final hook to the body-chamber with strong tubercles, tighter coiling, and presence of constrictions throughout.

Bostrychoceras from the Vistula Valley, Poland are described by Błaszkiewicz (1980) who recognized *B. polyplocum schlueteri* Błaszkiewicz, 1980 (p. 20, pl. 2, figs. 1, 4, 9–11) and *B.*

unituberculatum Błaszkiewicz, 1980 (p. 21, pl. 3, figs. 1–8; pl. 4, figs. 3–6) from the same locality, and *B. polyplocum polyplocum* from elsewhere, stating (p. 20) that it is the younger form. Subspecies *schluteri* is linked to *polyplocum* by transitions, has fifty ribs per whorl, and seems to be a synonym. The species *unituberculatum* has, as its name suggests, 'one row of tubercles in both early and late development stages . . . a character which differs . . . from all other taxons of the genus *Bostrychoceras*' (Błaszkiewicz 1980, p. 21). Specimens identical to the holotype (Błaszkiewicz 1980, pl. 3, figs. 3 and 6) occur in the Haldem material I have studied, and some specimens show a uni- and non-tuberculate stage at different points in development. I regard it as a further variant within a variable *polyplocum* although accept that there may be a temporal change in the range of variation within the species.

Several of the ill-preserved *Didymoceras* from the Upper Campanian of the Vistula Valley described by Błaszkiewicz seem to be based on variants of *polyplocum* as here conceived, as indicated in the synonymy.

C. (*C.*) *polyplocum zumayense* Wiedmann, 1962 (p. 200, pl. 9, fig. 5) is from the Lower Maastrichtian; it is described as having only forty ribs per whorl, but the holotype and only described specimen is part of a U-shaped body-chamber of a nostoceratid whose generic and subgeneric assignation are doubtful; it is best treated as a *nomen dubium*.

Occurrence. Upper Campanian. Northern Ireland, Norfolk, England, Assise P^3–Q in northern Aquitaine; Tercis (Landes), Torallola, Montesqui and elsewhere in Catalonia; Navarra, Spain; widespread in the Germanies, Poland, European Russia, Donbassa, Soviet Armenia, Transcaspia, Soviet Central Asia, Bulgaria, Iran, North Africa, Texas.

Family DIPLOMOCERATIDAE Spath, 1926, p. 81

[= Neocrioceratinae Spath, 1953, p. 17]

Subfamily DIPLOMOCERATINAE Spath, 1926, p. 81

[= Scalaritinae Ward, 1976, p. 455]

Genus NEOCRIOCERAS Spath, 1921, p. 51

Type species. *Crioceras spinigerum* Jimbo, 1894, p. 184, pl. 24, fig. 1, by the subsequent designation of Diener 1925, p. 192.

Subgenus SCHLUETERELLA Wiedmann, 1962, p. 205

Type species. *Ancyloceras pseudoarmatum* Schlüter, 1872*a*, p. 99, pl. 31, figs. 1–3, by original designation.

Neocrioceras (*Schlueterella*) ?*elegans* (Moberg, 1885)

Plate 17, figs. 3–6

1885 *Ancyloceras* ?*elegans* Moberg, p. 30, pl. 3, fig. 10.
1925 *Ancyloceras elegans* Moberg; Diener, p. 193.

Holotype. Moberg's original specimen (1885, pl. 3, fig. 10) from the Upper Campanian of Köpinge, Sweden.

Material. SP unregistered *ex* Arnaud Collection, two specimens: one from Assise P^3, St-Médard-de-Barbezieux (Charente); one from 'dallage à niveau de Peudry, P^3'.

Description. The smaller specimen is a slightly curved wholly septate fragment with a compressed oval whorl section (whorl breadth to height ratio 0·82). Delicate slightly rursiradiate ribs are weak across the dorsum but strengthen over the flanks. Pairs of ribs are linked at the ventrolateral shoulder by a blunt tubercle, the tubercles linked over the venter by a broad rib. These pairs are separated by two delicate, nontuberculate ribs, giving a rib index of 8. The larger fragment is worn, more strongly curved, with a whorl breadth to height ratio of 0·73. It shows the same style of ribbing and tuberculation as the first fragment, with a rib index of 7–8.

Suture with deeply incised bifid lobes and saddles.

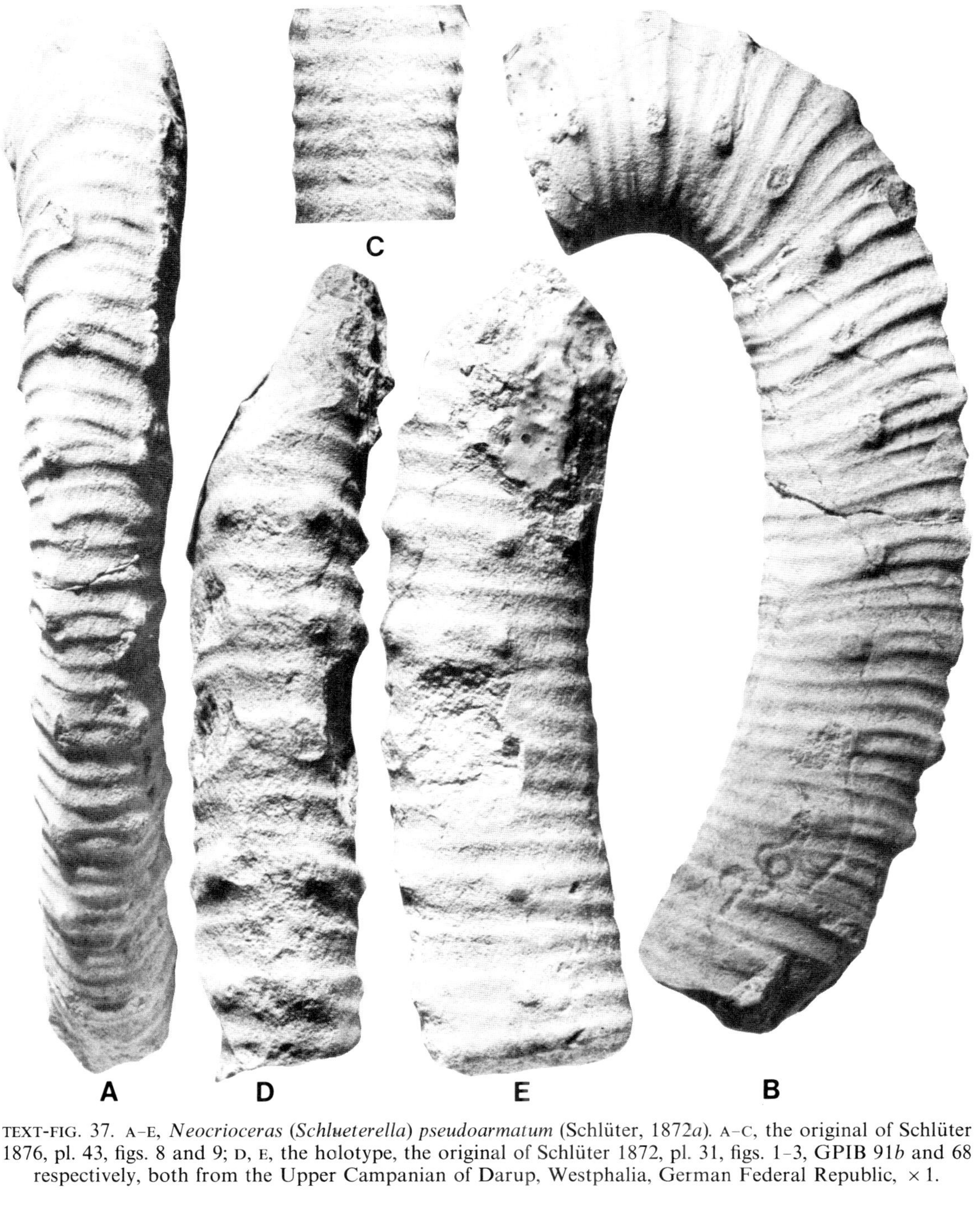

TEXT-FIG. 37. A–E, *Neocrioceras* (*Schlueterella*) *pseudoarmatum* (Schlüter, 1872*a*). A–C, the original of Schlüter 1876, pl. 43, figs. 8 and 9; D, E, the holotype, the original of Schlüter 1872, pl. 31, figs. 1–3, GPIB 91*b* and 68 respectively, both from the Upper Campanian of Darup, Westphalia, German Federal Republic, × 1.

Discussion. The holotype of *Ancyloceras elegans* is a battered fragment only with a rib index of 8 and an approximate whorl breadth to height ratio of 0·84 (measurements from figure). Ornament matches that of the present material, while they are broadly contemporaneous, supporting the determination. *Idiohamites*(?) *oronensis* Lewy, 1969 (p. 127, pl. 3, figs. 10 and 11) from the uppermost Mishash Formation of Israel, dated as Upper Campanian by its author, is a close ally, showing, however, ribs grouped in pairs as in the present species but with only a single nontuberculate rib between and pairs of ribs looping across the venter between tubercles; these differences may be within the limits of intraspecific variation, but with so little material one cannot be certain. *I.*(?) *circularis* Lewy, 1969 (p. 128, pl. 3, fig. 9; pl. 4, fig. 3) from the same horizon has a circular whorl section, similar flank ornament, but ribs that weaken or disappear on the venter. The suture is, however, on the same pattern as the present material.

This has been a neglected species, and while I am confident that the Swedish and French specimens represent the same species, the generic assignation is far less certain. *Schlueterella* generally have four rows of tubercles rather than two, and this criterion alone serves to distinguish *elegans* from the type and other species in our present state of knowledge. The described material of *N. pseudoarmatum* is, however, much larger than the present specimens (text-fig. 37), and the two may represent ontogenetic stages of the same genus.

Occurrence. Upper Campanian, Mucronatakritan of Köpinge, Sweden, Assise P^3 in Aquitaine.

Genus NEANCYLOCERAS Spath, 1926, p. 80

Type species. Ancyloceras bipunctatum Schlüter, 1872*a*, p. 98, pl. 29, figs. 1–3.

Neancyloceras cf. *bipunctatum* (Schlüter, 1872*a*)

Plate 16, fig. 5

1872*a* *Ancyloceras bipunctatum* Schlüter, p. 98, pl. 29, figs. 1–3.
1982 *Ancyloceras bipunctatum* Schlüter, 1872; Klinger, p. 221, figs. 1, 1–3, 2–8A–E, 9 (with full synonymy).

Lectotype. By the subsequent designation of Błaszkiewicz 1980, p. 29: the original of Schlüter 1872*a*, pl. 29, fig. 3, refigured by Klinger 1982, figs. 3 and 4*a*, from the Upper Campanian of Ahlten, German Federal Republic.

Material. SP unregistered *ex* Arnaud Collection, from Assise P^3, Bouteille (Dordogne).

Discussion. This scrap has an oval whorl section, deformed by crushing, a rib index of 4, with a suggestion of ill-preserved ventral tubercles, which are the basis for comparison with *Neancyloceras bipunctatum*, fully revised by Klinger (1982), to which the reader is referred for a discussion of the problems of the proper generic assignation of the species.

Occurrence. As for material.

EXPLANATION OF PLATE 16

Figs. 1–4, 6, 7. *Neoglyptoxoceras* (?) *retrorsum* (Schlüter, 1872*a*). 1–4, SP unregistered *ex* Arnaud Collection, from Assise P^3, Tauillard (Charente); 6 and 7, from 'Camp. inf. Livernant' (Charente).

Fig. 5. *Neancyloceras* cf. *bipunctatum* (Schlüter, 1872*a*). SP unregistered *ex* Arnaud Collection, from Assise P^3, Bouteille (Dordogne).

Figs. 8, 9, 12–17. *Trachyscaphites pulcherrimus* (Roemer, 1841). All SP unregistered *ex* Arnaud Collection. 8 and 9 from Assise QR, Courgeac-Beauvais; 12 and 13, from 'Meschers' [-sur-Gironde] (Charente-Maritime); 14 and 15, from Assise P^3, Le Caillaud (Charente-Maritime); 16 and 17, from Assise Q, Neuvic (Dordogne).

Figs. 10 and 11. *Pseudoxybeloceras* (*Parasolenoceras*) cf. *interruptum* (Schlüter, 1872*a*). SP unregistered *ex* Arnaud Collection, from Assise Q^2, Courgeac (Charente).

All figs. $\times 1$.

KENNEDY, *Neoglyptoxoceras*, *Neancyloceras*, *Trachyscaphites*, *Pseudoxybeloceras* (*Parasolenoceras*)

Genus NEOGLYPTOXOCERAS Collignon, 1969, p. 35
[? = *Epiglyptoxoceras* Collignon, 1969, p. 35]

Type species. *Neoglyptoxoceras magnificum* Collignon, 1969, p. 35, pl. 526, figs. 2074 and 2075; pl. 527, fig. 2077.

Diagnosis. Large, criocone, whorl section circular or compressed. Ornamented by strong, sharp, distant, commonly rursiradiate ribs. Suture with large auxiliary lobe in one (lateral?) saddle according to its author.

Discussion. These giant criocones differ from *Epiglyptoxoceras abnorme* Collignon, 1969 (p. 35, pl. 526, fig. 2076) in the weaker ornament and details of suture; it is probably a synonym. *Glyptoxoceras* Spath, 1925*b* (p. 30), with type species, by original designation, *Hamites rugatus* Forbes, 1846, p. 117, pl. 11, fig. 2 has been used for a wide range of generally fragmentary ribbed heteromorphs of the Santonian to Upper Maastrichtian. The type species and others described by Forbes (1846) are Lower Maastrichtian; the type species at least is known from fragments only with a compressed whorl section and loose open coiling. I refer the European *A. retrorsum* Schlüter, 1872*a* (p. 97, pl. 30, figs. 5–10) to *Neoglyptoxoceras* with a query; size and ribbing style are correct, but the sutures are unknown.

Occurrence. Campanian. Madagascar, France, German Federal Republic, Austria (Ultrahelvetic of the Gschliefgraben), USSR, Madagascar, and possibly Sweden.

Neoglyptoxoceras(?) *retrorsum* (Schlüter, 1872*a*)

Plate 16, figs. 1–4, 6, 7; plate 17, figs. 1 and 2; text-fig. 38

1872*a* *Ancyloceras retrorsum* Schlüter, p. 97, pl. 30, figs. 5–10.
?1885 *Anisoceras* (*Hamites*) *crispatus* Moberg, p. 30, pl. 3, fig. 10.
1889 *Ancyloceras retrorsum* Schlüter; Griepenkerl, p. 105.
1905 *Ancyloceras retrorsum* Schlüter; Wegner, p. 210.
1913 *Ancyloceras retrorsum* Schlüter; Nowak, p. 383, pl. 40, figs. 1 and 5.
1925 *Ancyloceras retrorsum* Schlüter; Diener, p. 194.
1925*b* *Glyptoxoceras retrorsum* Schlüter; Spath, p. 31.
1951 *Ancyloceras retrorsum* Schlüter; Mikhailov, p. 87, pl. 5, fig. 32; pl. 13, fig. 56.
1959 *Neancyloceras retrorsum* (Schlüter); Naidin and Shimanskij, p. 182, pl. 3, fig. 7.
1976 *Glyptoxoceras retrorsum* (Schlüter); Atabekian and Khakhimov, p. 61, pl. 10, fig. 3; pl. 11, fig. 1.
1980 *Glyptoxoceras retrorsum* (Schlüter, 1872); Błaszkiewicz, p. 28, pl. 55, figs. 3–5.
1984 *Neoglyptoxoceras*? cf. *retrorsum* (Schlüter, 1872): Kennedy and Summesberger, p. 168, pl. 6, fig. 1.

Types. Lectotype, by the subsequent designation of Spath 1925*b*, p. 31 (footnote), GPIB 67, the original of Schlüter 1872*a*, pl. 30, figs. 5–7, from the Upper Campanian of Coesfeld, Westphalia, German Federal Republic (text-fig. 38A–C). The smaller figured paralectotype of Schlüter 1872*a*, pl. 30, fig. 8, GPIB 67*b*, is from the same horizon and locality (text-fig. 38D).

Material. SP unregistered *ex* Arnaud Collection: two specimens from Assise P^3 at Tauillard (Charente); one specimen from the 'Camp. inf. Livernant' (Charente).

Description. All specimens are composite internal moulds of curved fragments, with a maximum whorl height of 16 mm. The whorl section is a compressed oval with a whorl breadth to height ratio 0·65–0·8, although the material is crushed. The rib index is 6–8. The ribs are strong, narrow, straight, and rursiradiate on the flank, strong on the venter, but effaced on the dorsum.

Sutures not seen.

Discussion. These curved fragments show few diagnostic features, yet closely match the German material, notably the smaller figured paralectotype (text-fig. 38D). Most individuals referred to this species are curved fragments, suggesting an open criocone coiling. *Anisoceras* (*Hamites*) *crispatus* Moberg, 1885 (p. 30, pl. 3, fig. 10) has a more compressed whorl section but this is clearly due to post-morten crushing. It has a rib index of 7–8 and is here regarded as a possible synonym only because it is significantly older than the types of *retrorsum* (from Moberg's *granulatus* Zone). It corresponds closely to the Lower Campanian specimen from Livernant (Pl. 16, figs. 6 and 7).

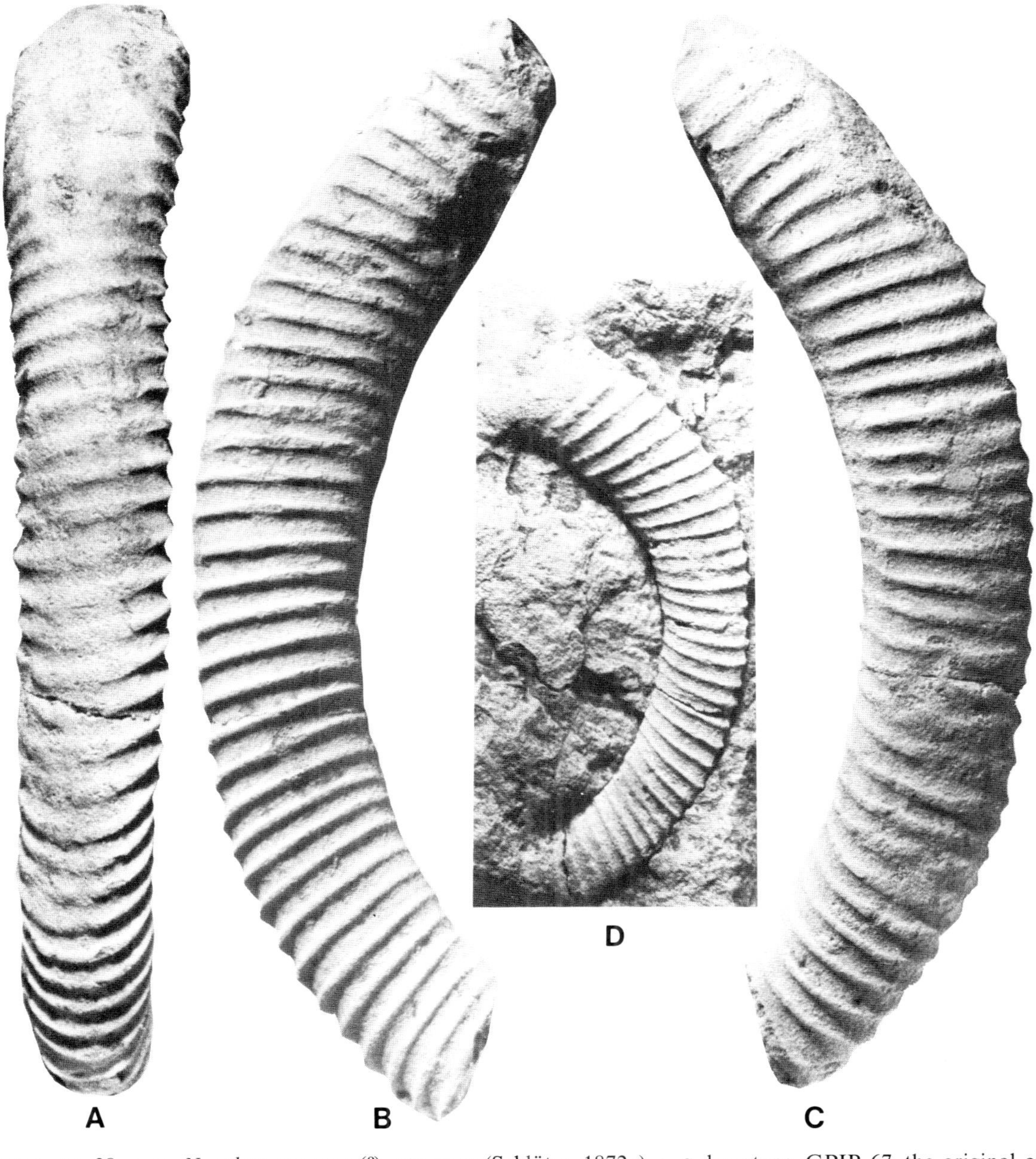

TEXT-FIG. 38. A–D, *Neoglyptoxoceras* (?) *retrorsum* (Schlüter, 1872*a*). A–C, lectotype, GPIB 67, the original of Schlüter 1872*a*, pl. 30, figs. 5–7; D, paralectotype, GPIB 67a, the original of Schlüter's pl. 30, fig. 8; both specimens from the Upper Campanian of Coesfeld, Westphalia, German Federal Republic, ×1.

Occurrence. Upper and Lower Campanian of northern Aquitaine. Upper Campanian of Westphalia, German Federal Republic, Austria (Ultrahelvetic, Gschliefgraben); Lower Campanian of Poland; Upper Campanian of European Russia, Ukraine, Caucasus, Soviet Armenia, Kopet Dag, and possibly Sweden.

Subfamily POLYPTYCHOCERATINAE Matsumoto, 1938, p. 193
[*nom. transl.* Wiedmann, 1962, p. 185 *ex* Polyptychoceratidae Matsumoto].

Genus PSEUDOXYBELOCERAS Wright and Matsumoto, 1954, p. 119

Type species. Hamites quadrispinosus Jimbo, 1894, p. 185 (39), pl. 23 (7), figs. 3 and 4, by original designation.

Subgenus PARASOLENOCERAS Collignon, 1969, p. 44

Type species. Parasolenoceras splendens Collignon, 1969, p. 44, pl. 530, figs. 2087 and 2088.

Pseudoxybeloceras (*Parasolenoceras*) cf. *interruptum* (Schlüter, 1872*a*)

Plate 16, figs. 10 and 11

1872*a* *Hamites interruptus* Schlüter, p. 102, pl. 32, figs. 8 and 9.
1982 *Pseudoxybeloceras* (*Parasolenoceras*) *interruptus* (Schlüter, 1872); Klinger, p. 237, text-fig. 8F, G.
1984 *Pseudoxybeloceras* (*Parasolenoceras*) *interruptum* (Schlüter, 1872); Kennedy and Summesberger, p. 167, pl. 9, figs. 5, 10, 11 (with full synonymy).

Type. Holotype, by monotypy, the original of Schlüter 1872*a*, pl. 32, figs. 8 and 9, Göttingen University Collection no. 65-13, from the Upper Campanian of Ahlten, German Federal Republic (refigured by Klinger 1982, text-fig. 8F, G).

Material. SP unregistered *ex* Arnaud Collection, from Assise Q^2 at Courgeac (Charente).

Discussion. This rather poorly preserved fragment shows ptychoceratid coiling and coarse straight prorsiradiate ribs (rib index is 3–4) with ventral tubercles between which the ribs appear to weaken. It closely resembles the holotype.

Occurrence. As for material. *Pseudoxybeloceras* (*Parasolenoceras*) *interruptum* is an Upper Campanian *polyplocum* Zone species in Ireland, the German Federal Republic, Austria (Ultrahelvetic, Gschliefgraben), Donbass, and southern USSR.

Family BACULITIDAE Gill, 1871, p. 3
[= Eubaculitinae Brunnschweiler, 1966, p. 24]

Genus BACULITES Lamarck, 1799, p. 80
[= *Homaloceratites* Hupsch, 1768, p. 110 (*non binomen*); *Euhomaloceras* Spath, 1926, p. 80]

Type species. Baculites vertebralis Lamarck, 1801, p. 103, by subsequent designation of Meek 1876*a*, p. 391.

EXPLANATION OF PLATE 17

Figs. 1 and 2. *Neoglyptoxoceras* (?) *retrorsum* (Schlüter, 1872*a*). SP unregistered *ex* Arnaud Collection, from Assise P^3, Tauillard (Charente).
Figs. 3–6. *Neocrioceras* (*Schlueterella*) ? *elegans* (Moberg, 1885). SP unregistered *ex* Arnaud Collection, from Assise P^3, 'dallage à niveau de Peudry' and St Médard-de-Barbezieux (Charente) respectively.
Figs. 7–9, 13–15, 21–23. *Baculites* sp. 1. All SP unregistered *ex* Arnaud Collection; 7, from St Hilaire [-de-Montmoreau] (Charente); 13–15, from Assise P^1, Montmoreau (Charente); 21–23, from Assise P^3, St.-Médard-de-Barbezieux (Charente).
Fig. 10. *Baculites* sp. 4. SP unregistered *ex* Arnaud Collection, from Assise P^2, St Hilaire [-de-Montmoreau] (Charente).
Figs. 11, 12, 16–20. *Baculites* sp. 2. All SP unregistered *ex* Arnaud Collection. 11, 12, 16, 17, from Assise P^1, Livernant (Charente); 18–20, from Assise P^2, Montmoreau, Route de Girat (Charente).
All figs. ×1.

KENNEDY, *Neoglyptoxoceras*, *Neocrioceras* (*Schlueterella*), *Baculites*

Discussion. See Kennedy (1984*a*) for a recent diagnosis and discussion of the genus. I have recently clarified the limits of the classic Coniacian (1984*a*) and Maastrichtian species (1986 in press *b*) of *Baculites* described from Europe. The nomenclature of the Campanian forms, in contrast, is in a state of confusion. Most authors either ignore the baculitids of this stage (e.g. de Grossouvre 1894; Błaszkiewicz 1980) or use names of older or younger forms such as *incurvatus* Dujardin, 1837, *anceps* Lamarck, 1822, or *vertebralis* Lamarck, 1801. *Baculites* is the commonest ammonite in the Aquitaine Campanian, and the only one readily found in the field these days. In spite of this the material is poor and fragmentary and I hesitate to introduce new names at this time, recognizing five forms, left in open nomenclature. The Lower Maastrichtian yields three distinctive species, one of which can be attached to *leopoliensis* Nowak, 1908; the others I again leave in open nomenclature. I have no doubt that *Baculites* are of great potential stratigraphic value in the European Campanian, when better material is published. In the US Western Interior no less than eleven of the seventeen to eighteen Campanian zones have *Baculites* as index species.

Baculites sp. 1

Plate 17, figs. 7–9, 13–15, 21–23; Plate 18, figs. 18–22; Plate 23, figs. 1 and 7; text-fig. 8A, C.

1888 *Baculites incurvatus* Duj.; Holzapfel, p. 64 (*pars*), pl. 4, figs. 5 and 6.
1906 *Baculites incurvatus* Dujardin; Müller and Wollemann, p. 4, pl. 2, figs. 2–5.

Material. SP unregistered *ex* Arnaud Collection: 2 specimens from Assise P^2, Montmoreau (Charente); 5 from Assise P^2, St Hilaire (Charente); 1 specimen from Assise P^2, La Caillerie (Charente); 1 doubtful specimen from Assise P^3, St-Médard-de-Barbezieux (Charente); 1 doubtful specimen from Tauillard (Charente). OUM KZ13957 from the Upper Campanian of Les Ratoux, 1·5 km east of St. Victor (Dordogne) may also belong here.

Description. All specimens are internal moulds. Whorls slowly expanding, compressed, with breadth to height ratio 0·65. Intercostal whorl section oval. Dorsum broadly rounded, flanks flattened, converging to venter only slightly more narrowly rounded than dorsum. Ornament dominated by coarse, distant transverse crescentic dorsolateral bullae which give rise to a strongly prorsiradiate rib that sweeps forwards and effaces on the outer flank; the angle of intersection with the line of the venter is 30–40° and the bullae are separated by a distance equal to 1·7 to 2·4 major whorl diameters. Ornament of interspaces between major bullae varies between specimens. Some appear smooth, or bear concave growth lines, sometimes strengthened into finer riblets, up to three of which may develop in a single interspace (Pl. 23, fig. 1), producing a distinctive serrated venter; another specimen lacks obvious riblets on the flank but nevertheless shows ventral serrations.

Suture moderately incised for genus (text-fig. 8A, C).

Discussion. The distantly nodate specimens with smooth interspaces correspond to what Holzapfel (1888, p. 64, pl. 4, figs. 5 and 6) and Müller and Wollemann (1906, p. 4, pl. 2, figs. 2–5) called *Baculites incurvatus.* Dujardin's species (1837, p. 232, pl. 17, fig. 13; see Kennedy 1984*a*, p. 143, pl. 32, figs. 12, 15–19; pl. 33, figs. 1–22; text-figs. 41 and 42F–M) is significantly smaller, has an ovoid rather than near oval intercostal whorl section, the greatest breadth close to the dorsum rather than just below mid-flank, weaker bullae, separated by a distance equal to or only slightly greater than the whorl height. Adult body-chambers also show a marked curvature. The main occurrence of this species is Coniacian–Santonian.

EXPLANATION OF PLATE 18

Figs. 1, 4–12. *Baculites leopoliensis* Nowak, 1908. All SP unregistered *ex* Arnaud Collection, from Assise R^2, Maurens (Dordogne).
Figs. 2 and 3. *Baculites* sp. 5. SP unregistered *ex* Arnaud Collection from Assise P^3, Courgeac (Charente).
Figs. 13 and 14. *Baculites* sp. 2. SP unregistered *ex* Arnaud Collection, from Assise P^1, Livernant (Dordogne).
Figs. 15–17. *Baculites* sp. 4. SP unregistered *ex* Arnaud Collection, from Assise P^2, Montmoreau (Charente).
Figs. 18–22. *Baculites* sp. 1. SP unregistered *ex* Arnaud Collection. 18, 19, 22, from St. Hilaire [-de-Montmoreau] (Charente); 20 and 21, from Montmoreau (Charente).
All figs. × 1.

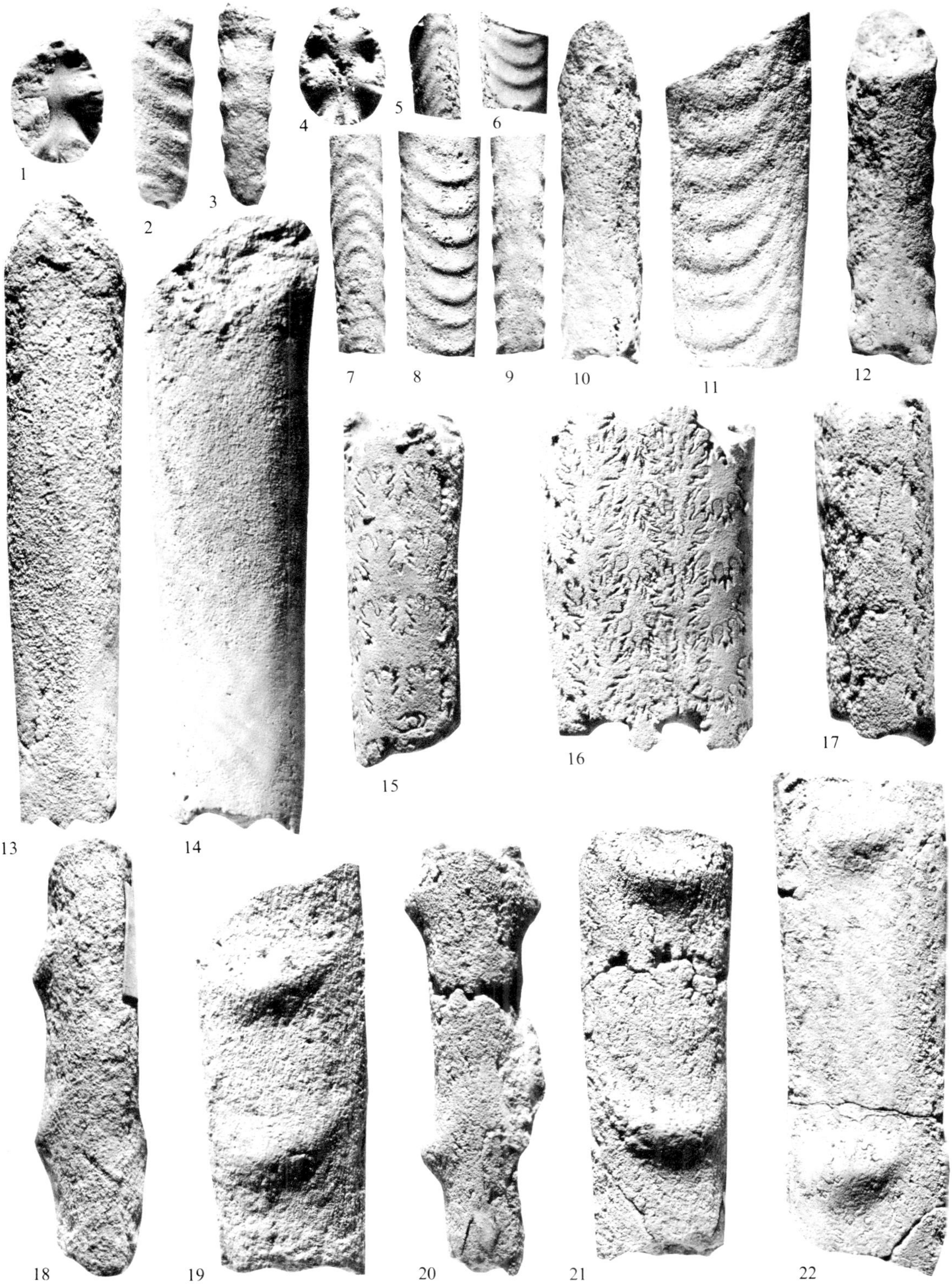

KENNEDY, *Baculites*

Nodose baculitids are abundant in the Upper Cretacous of the Indo-Pacific region. They show wide intraspecific variation (e.g. Kennedy *et al.* 1974; Klinger and Kennedy 1977). *B. capensis* Woods, 1906 (p. 342, pl. 44, figs. 6 and 7; see Klinger and Kennedy 1977, p. 71, figs. 2A–F and 3G) is an important Santonian species; it has closer-spaced conical rather than crescentic tubercles. *B. boulei* Collignon, 1931 (p. 35, pl. 5, fig. 2; pl. 9, fig. 14) (see Matsumoto 1959*b*, p. 119, pl. 32, fig. 7; pl. 33, figs. 4–7; text-figs. 27*a*, *b*, 28–32) has a stouter whorl section and closer-spaced tubercles; the whorl section (Matsumoto 1959*b*, text-figs. 28–32) shows them to be close to the dorsum with maximum width there rather than near mid-flank, as in the present species. The very wide spacing of tubercles in the present form alone suffices to separate it from most of the diverse forms described by Collignon (1969, 1970) while they show dorsal displacement of tubercles compared to the present species. Decision as to whether these numerous Madagascan forms are truly distinct must await variation studies on Collignon's material. The closest Madagascan species appears to be *B. sparsinodosus* Collignon, 1969 (p. 23, pl. 521, figs. 2052–2054) but here too the whorl section differs, with the tubercles dorsally displaced compared to the present material.

B. vanhoepeni Venzo, 1936 (p. 116, pl. 10 (6), figs. 11 and 12 (as *B. vagina* Forbes var. *van hoepeni*)) (see Klinger and Kennedy 1977, p. 73, figs. 2G, K, 3A, H, I, 4A–C, 5C) is a Middle Campanian species from South Africa that has very distant ribbing and, in some variants, very distant crescentic tubercles (Klinger and Kennedy 1977, fig. 4A–C); it is the closest described species to the present material.

Some Aquitaine specimens are labelled *B. distans* in Arnaud's hand; this appropriate name is, however, a *nomen nudum* published, so far as I know, only by de Grossouvre in a list (1901, table 35).

Occurrence. As for material; Lower Campanian Vaals Formation on The Netherlands–Belgian border; Lower Campanian of Braunschweig, German Federal Republic.

Baculites sp. 2

Plate 17, figs. 11 and 12, 16–20; Plate 18, figs. 13 and 14; Plate 23, figs. 28 and 29

1888 *Baculites* (?)*incurvatus* Duj.; Holzapfel, p. 64 (*pars*), pl. 5, fig. 10.

Material. SP unregistered *ex* Arnaud Collection: two specimens from Assise P^1, Livernant (Charente); one from Assise P^2, Montmoreau (Charente); one labelled 'P^2, Montmoreau, route de Girat' (Charente); one from P^2, La Côte, Juillaguet (Charente).

Description. Slowly expanding, whorl section compressed (whorl breadth to height ratio 0·63–0·66), oval with broadly rounded dorsum, flattened flanks and venter only slightly more narrowly rounded than dorsum; greatest breadth dorsal of mid-flank. Shell surface of internal mould smooth or with faint ventral notches. Composite mould shows distinctive growth lines, riblets, and striae. These delineate the apertural outline and are deeply and asymmetrically concave, intersecting the line of the dorsum at 37° (Pl. 17, fig. 19).

Suture moderately incised for genus (text-fig. 8F, G).

Discussion. Cross-section and suture are identical to those of *Baculites* sp. 1, described above, while the course of the growth lines is similar. It may well be that they represent no more than the smooth variants of the same species, for both European *B. anceps* Lamarck, 1822 (p. 648; see Kennedy 1986*a*, p. 58, pl. 11, figs. 12–14; pl. 12, figs. 7–11; text-figs. 3E–H and 7A–C) and *B. vertebralis* Lamarck, 1801 (Kennedy, 1986*a*, p. 57, pl. 11, figs. 6–11; pl. 12, figs. 1–6; text-figs. 3A–D, 7D–F, 8) of the Maastrichtian include smooth and ribbed variants, as do the various Campanian species found in the US Western Interior (Scott and Cobban 1965; see also Kennedy and Cobban 1976, figs. 15 and 16).

Smooth European Campanian *Baculites* have generally been referred to *B. vertebralis* or some other Maastrichtian species (e.g. Schlüter 1876, p. 143, pl. 39, fig. 11); I suspect some belong to the present form. Holzapfel (1888, pl. 5, fig. 10) figured the present form as *B.* (?)*incurvatus* from the Vaals Formation; the Collections of the Institut Royal des Sciences Naturelles de Belgique contains many comparable specimens from this Lower Campanian unit, associated with the distantly nodate form, as in Aquitaine.

Occurrence. Lower Campanian, Assise P^1–P^2 in northern Aquitaine: Vaals Formation of Limburg Province, The Netherlands, Liège Province, Belgium, and the German Federal Republic.

Baculites sp. 3

Plate 23, figs. 11–13, 16–18, 20 and 21

Material. SP unregistered *ex* Arnaud Collection: 2 specimens from Assise Q^2, Puyvigier (Charente); 1 from Assise Q, Rioux Martin (Charente); 1 from Assise Q^2, Aubeterre-sur-Dronne (Charente); 4 from Assise Q, Issac (Dordogne); 1 from Assise Q^2, Neuvic (Dordogne). OUM KZ13925–13931 from Assise Q, Brossac (Gare) (Charente).

Description. Slowly expanding, whorl section compressed ovoid-pyriform with whorl breadth to height ratio 0·7–0·75, greatest breadth dorsal to mid-flank. Venter narrowly rounded to fastigiate but not acute, dorsum broadly rounded. Flanks ornamented by deep crescentic grooves that demarcate broad, scale-like ridges. These have a steep adapical face (where the constrictions are deepest) and a gently inclined adapertural face. The constrictions cross the venter in a broad convexity, sweep back over the dorsolateral shoulder and are concave on the flank, projecting strongly forwards to the venter. Constrictions vary from regular development at a spacing such that one rib and parts of two constrictions occur in a distance equal to the whorl height in some specimens while in others they are widely and irregularly spaced. A few non-constricted fragments with comparable whorl section and growth lines are also placed here.

Sutures not seen.

Discussion. The material is poorly preserved, but is distinct from all other described European Campanian–Maastrichtian *Baculites*. Some are labelled *B. anceps* in Arnaud's hand, but although constricted examples of this Upper Maastrichtian species may appear similar in side view (e.g. Howarth 1965, pl. 6, figs. 2, 4, 5) they have a highly distinctive whorl section with acute venter (Howarth 1965, text-figs. 5–12; Kennedy 1986*a*, text-fig. 3E–H). There are similarities to the Lower Campanian *B. sulcatus* Baily, 1855 (p. 457, pl. 11, fig. 5*c* only; see Klinger and Kennedy 1977, p. 75, fig. 3B–E, J–L), but this has narrower ribs and constrictions, both of which continue over the venter in distinctive fashion not seen in the Aquitaine specimens. Another species with similar type of ornament in *B. columna* Morton, 1834 (p. 44, pl. 19, fig. 8) (see Stephenson 1941, p. 405, pl. 76, figs. 1–4) but the ribs are much narrower than the very broad interspaces/constrictions.

Occurrence. As for material.

Baculites sp. 4

Plate 17, fig. 10; Plate 18, figs. 15–17; text-fig. 8H

Material. SP unregistered *ex* Arnaud Collection: 1 specimen from Assise P^2, Montmoreau (Charente).

Description. Slowly expanding with very compressed slightly ovoid whorl section. Whorl breadth to height ratio 0·53–0·55 with greatest breadth dorsal of mid-flank. Dorsum broadly rounded, flanks flattened, convergent, venter more narrowly rounded than dorsum. Surface of mould smooth or with a trace of strongly concave growth lines and striae.

Suture moderately incised for genus (text-fig. 8H).

Discussion. These fragments are much more compressed than the other Lower Campanian *Baculites* from Aquitaine and may represent a separate species.

Occurrence. As for material.

Baculites sp. 5

Plate 18, figs. 2 and 3

Material. SP unregistered *ex* Arnaud Collection, one specimen from Assise P^3, Courgeac (Charente).

Description. Specimen is a slightly distorted composite mould. Whorl section appears to have been ovoid with broad, flattened dorsum. Rib index 2. Broad convex ribs on the dorsum strengthen over the dorsolateral shoulder into coarse crescentic concave flank ribs, only slightly narrower than the interspaces, sweeping forwards over the ventrolateral region, and crossing the venter in a narrow convexity.

Sutures not seen.

Discussion. The close coarse ribbing of this specimen, persistent over dorsum and venter, distinguishes it from all other Aquitaine *Baculites.* It recalls *B. columna* Morton, 1834 (p. 44, pl. 19, fig. 8) but the flank ribs are straight rather than concave in that species.

Occurrence. As for material.

Baculites sp. 6

Plate 23, figs. 24–26

Material. OUM K19878, 199880, 19891, and numerous fragments from Assise R of Arnaud in road cuttings at Maurens-le-Bost (Dordogne). Dr. M. Neumann (Paris) places this section in foraminifer Biozone CVIII, equivalent to Bed 9 or 10 of the Maurens borehole. OUM K213863, from the lumachelle facies of Assise R of Arnaud, roadside ditches between La Petite Metairie and railway crossing, minor road by Chez Vardon, 2·2 km south of Baignes-Ste-Radegorde (Charente-Maritime): block with several fragments. Dr. M. Neumann (Paris) has examined the matrix of this material; it contains *Abrardia mosae*, indicating foraminifer Biozone CVII.

Description. Small, slowly expanding compressed with ovoid whorl section, whorl breadth to height ratio 0.75. Venter only slightly more narrowly rounded than dorsum; flanks flattened. Internal and external moulds smooth.

Sutures not seen.

Discussion. Whorl section and lack of ornament separate this species from all others found in Aquitaine. Of other European species it most closely resembles *Baculites vertebralis* Lamarck, 1801 (p. 103, see Kennedy 1986*a*, p. 57, pl. 11, figs. 6–11; pl. 12, figs. 1–6; text-figs. 3A–D, 7D–F, 8) which also has an oval whorl section. Biometric analysis of a large sample ($n = 50$) of individuals from the Upper Maastrichtian Calcaire Kunraed of Kunraed, Limburg, The Netherlands (Kennedy in press *b*) shows this to have a whorl breadth to height ratio 0·8 to 0·4, with the majority of specimens more compressed than the present material. *B. vertebralis* includes ribbed and constricted forms, features not seen in the present, albeit small, collection.

Occurrence. As for material.

Baculites leopoliensis Nowak, 1908

Plate 18, figs. 1, 4–12

1908 *Baculites anceps* Lam. sp. em. Nowak 1. varietas *Leopoliensis* Nowak, p. 328 (*pars*), pl. 14, figs. 1–5, 10, 11; text-figs. 1–5 on p. 329; ? text-figs. 5–10 on p. 311.
1925 *Baculites anceps-leopoliensis* Nowak; Diener, p. 57.
non 1951 *Baculites anceps* Lam. var. *leopoliensis* Nowak; Mikhailov, p. 46, pl. 3, fig. 15.
non 1964 *Baculites anceps leopoliensis* Nowak; Tzankov, p. 149, pl. 10, fig. 2.
non 1974 *Baculites anceps leopoliensis* Nowak, 1908; Naidin, p. 164, pl. 53, fig. 5.
non 1976 *Baculites anceps leopoliensis* Nowak; Atabekian and Khakhimov, p. 96, pl. 1, figs. 11–13.
1986*c* *Baculites leopoliensis* Nowak, 1908; Kennedy, p. 1013, pl. 2, figs. 1, 2, 11, 12; pl. 3, figs. 22–24.

Lectotype. By the subsequent designation of Kennedy 1986*c*, p. 1013, the original of Nowak 1908, pl. 14, fig. 1 from Lipnicki, formerly in Galicia, but now in the Ukrainian SSR.

Material. SP unregistered *ex* Arnaud Collection: nine specimens from Assise R^2, Maurens (Dordogne).

Description. All specimens are internal or external moulds. Slowly expanding, compressed, whorl breadth to height ratio 0·70–0·78, the greatest breadth dorsal of mid-flank. The dorsum is broadly rounded, the flanks flattened, converging to narrowly rounded venter.

Low broad ribs, 3 to 3·5 in a distance equal to the whorl height, cross the dorsum in a broad convexity. They strengthen markedly at the dorsolateral shoulder into a transverse crescentic rib that occupies the dorsal two thirds of the flank, sweeping forwards and attenuating over the outer third before flexing back to cross the venter in a narrow convexity. The crescentic ribs sometimes bifurcate over the venter or short intercalatories are inserted, with five to six ventral ribs in a distance equal to the whorl height.

Sutures not seen.

Discussion. The ribbing pattern of these specimens is quite distinct from all other described European Campanian–Maastrichtian *Baculites* species. It compares closely, however, with small distorted specimens from the niveau de transition between the Marnes de Plagne and Calcaire nankin in the Petites-Pyrénées (Kennedy 1986*c*), which link to the large specimens from the same area that agree with the lectotype. True *B. anceps* Lamarck, 1822 (p. 648) (see Kennedy 1986*a*, p. 58, pl. 11, figs. 12–14; pl. 12, figs. 7–11; text-figs. 3E–H and 7A–C) has a distinctive whorl section, with a sharp venter and although both ribbed and smooth forms are known, none of the former show the multiplication of ribs on the venter. *B. vertebralis* Lamarck, 1801 (p. 103; see Kennedy 1986*a*, p. 57, pl. 11, figs. 6–11; pl. 12, figs. 1–6; text-figs. 3A–D, 7D–F, 8) even when ribbed has an oval whorl section with broadly rounded dorsum and venter. *B. leopoliensis* has been confused with *B. knorrianus* Desmarest, 1817 (p. 48, pl. 1, fig. 3); both Nowak (1908) and Mikhailov (1951) place the latter in the synonymy of the former, although *knorrianus* was validly introduced. The distinctive whorl section shown in the figure of Knorr can be matched in specimens illustrated by Favre (1869, p. 27, pl. 7, fig. 2) and numerous specimens in the BMNH, MMK, and NHMW Collections. These show it to be a large compressed, smooth to feebly ornamented Lower Maastrichtian species that never develops the distinctive ribbing of *leopoliensis*.

Occurrence. Lower Maastrichtian of the environs of Lvov, Ukrainian SSR (formerly Lemberg, Galicia). Niveau de transition of the Petites-Pyrénées, where it is also Lower Maastrichtian. The Maurens specimens, labelled *anceps*, are listed by Arnaud (1892) from his Bed *h*, corresponding to Bed 3 of Neumann, *et al.* (1984). Madame M. Neumann kindly examined the matrix of a specimen and confirms this as the horizon of the material.

Baculites sp. 7

Plate 23, figs. 4–6, 8–10, 14, 15, 19, 22, 23, 27

Material. SP *ex* Arnaud Collection: numerous specimens from Assise R of Arnaud, the Silex de Mussidan of Mussidan (Dordogne).

Description. Slowly expanding with compressed whorl section, whorl breadth to height ratio 0·7 with greatest breadth just dorsal of mid-flank. Dorsum broadly, flanks slightly rounded, venter narrowly rounded. Ornament consists of blunt, broad, crescentic flank ribs, two to three in a distance equal to the whorl height. They are transverse on the dorsal two-thirds of the flank and sweep forwards over the outer third, but disappear well before the line of the venter.

Sutures not seen.

Discussion. This species occurs in the same slabs as *Brahmaites brahma* (Forbes, 1846). It is a very generalized *Baculites* indeed, easily distinguished from *B. leopoliensis*, described above, by the absence of intercalated ribs and failure of all ribs to cross the venter. One specimen (Pl. 23, fig. 27) shows a slightly higher rib density, linking it to *leopoliensis*, although the ventral ribs are absent. It may well be a precursor of *B. anceps* itself, ribbed variants of which may have not dissimilar ornament (e.g. Howarth 1965, pl. 4, fig. 4). The ribs of the present form do not pass over the venter, however, none show constrictions and the whorl section is ovoid rather than pyriform with sharp venter as in Lamarck's species.

Occurrence. Lower Maastrichtian, Silex de Mussidan of Mussidan (Dordogne).

Superfamily SCAPHITACEAE Gill, 1871, p. 3

[*nom. transl.* Wright and Wright, 1951, p. 13 *ex* Scaphididae Gill 1871, p. 3]

Family SCAPHITIDAE Gill, 1871, p. 3

Subfamily SCAPHITINAE Gill, 1871, p. 3

[*nom. transl.* Wright, 1953, p. 473 *ex* Scaphitidae Gill, p. 3]

Genus SCAPHITES Parkinson, 1811, p. 145

Type species. Scaphites equalis. J. Sowerby, 1813, p. 53, pl. 18, figs. 1–3.

Subgenus SCAPHITES (SCAPHITES) Parkinson, 1811, pl. 45

[= *Anascaphites* Hyatt, 1900, p. 572; *Jahnnites* Hyatt, 1900, p. 572; *Holcoscaphites* Nowak, 1911, p. 564]

Scaphites (*Scaphites*) *haugi* de Grossouvre, 1894

Plate 21, figs. 12–16

1894 *Scaphites haugi* A. de Grossouvre sp., p. 244, pl. 31, fig. 5.
1925 *Scaphites haugi* de Grossouvre; Diener, p. 199.
1928 *Discoscaphites ?haugi* de Grossouvre; Reeside, p. 30.
1970 *Scaphites* cf. *haugi* de Grossouvre; Atabekian and Akopian, p. 39, pl. 2, figs. 3 and 4.

Holotype. By monotypy, the original of de Grossouvre 1894, pl. 31, fig. 5, SP unregistered *ex* Arnaud Collection, from Assise P^3, Tauillard (Charente).

Material. A topotype, SP unregistered *ex* Arnaud Collection; no. 83421 in the J. P. Platel Collection from that author's Campanian 3 near Périgueux (Dordogne).

Description. The holotype is a macroconch 61 mm long. It and the other material are somewhat distorted composite moulds. Spire very involute with a tiny umbilicus. Whorl section subquadrate in costal section with broadly rounded flanks and broadly rounded venter. Twelve rather coarse, distant straight feebly prorsiradiate primary ribs arise alternately at or just outside umbilical shoulder and strengthen markedly across flanks, terminating in a strong conical ventrolateral tubercle at juncture of flank and venter. The broadly rounded venter is crossed by narrow sharp secondary ribs looped in pairs between tubercles with a single, equal intercalatory between. Shaft straight, partially occluding umbilicus. Ribs coarsen, then efface. There is a strong conical umbilical tubercle at the beginning of the shaft and a second, massive bulge at the beginning of the final hook. Faint, low, broad swellings extend from these to the seven massive ventrolateral clavi. There are traces of fine ribs linking clavi across the venter at the beginning of the shaft and on the final hook but they are absent, possibly due to wear, over most of the length.

Sutures not seen.

Discussion. De Grossouvre compared this species to *Scaphites binodosus* Roemer, 1841 (p. 90, pl. 13, fig. 6). All specimens referred to this species are microconchs, and I here designate the specimen in the Roemer Collection figured by Frech (1915, fig. 6) lectotype. It survives in the Collections of the Geological Museum, Wrocław, Poland and I reproduce here a photograph of the specimen, supplied by Dr J. Gorczya-Scala (text-fig. 39C). It is from the Lower Campanian of Dülmen, Westphalia. At this locality and others the species co-occurs with *S. inflatus* Roemer, 1841 (p. 90, pl. 14, fig. 3), the lectotype, here designated being Roemer's original figured specimen, now lost, and from Dülmen, Westphalia, illustrated by French (1915, fig. 8). All figured *inflatus* are macroconchs; *inflatus* and *binodosus* have identical phragmocones but differ in that *binodosus* develops umbilical tubercles on the body-chamber whereas they are absent in *inflatus*. Unpublished work by C. W. Wright and myself leads to the conclusion that they are a dimorphic pair; I here select *binodosus* rather than *inflatus* as the name for the taxon because of its wider usage and the existence of a type specimen. Macroconch *binodosus* (text-fig. 39D, E) differ from *haugi* in their much greater size—a specimen from Dülmen is at least 120 mm long (GPIB *ex* Schlüter Coll., unregistered), the presence of distinct narrow flank ribs and lack of umbilical tubercles. Microconch *binodosus* (text-fig. 39A–C) have a quite different shell form, numerous umbilical tubercles and distinctive broad, flat flank ribs. The phragmocones of the two species are distinguished by the much coarser distant flank ribs and coarser tubercles of *haugi*.

S. hippocrepis (DeKay, 1828), the commonest *Scaphites* in Aquitaine, is much smaller. The spire has a rounded whorl section with much weaker or no tubercles while the body-chamber is much more weakly tuberculate with, in some specimens, dense flank ribs and a lateral tubercle (Pl. 20, figs. 4, 10, 13, 16).

Occurrence. As for material. Atabekian and Akopian (1970) record *S.* cf. *haugi* from the Campanian of Soviet Armenia.

TEXT-FIG. 39. A–E, *Scaphites* (*Scaphites*) *binodosus* (Roemer, 1841). A, B, a microconch, GPIB 60*a*, the original of Schlüter 1872*a*, pl. 24, figs. 4–6; C, the lectotype, Geological Museum, Wroclaw, Poland, the original of Roemer 1841, pl. 13, fig. 6 (print supplied by Dr J. Gorczya-Scala); D, E, macroconch, GPIB 59*a*, the original of Schlüter 1872*a*, pl. 24, figs. 1–3; all specimens from the Lower Campanian of Dülmen, Westphalia, ×1.

Scaphites (*Scaphites*) *hippocrepis* (DeKay, 1828)

Plate 13, fig. 5; Plate 19, figs. 1–34; Plate 20, figs. 1–6, 10–19; Plate 21, figs. 1–11; Plate 23, figs. 2 and 3

1828 *Scaphites hippocrepis* DeKay, p. 273, pl. 5, fig. 5.
1859 *Scaphites nanclasi* Coquand, p. 996.
1860 *Scaphites Nanclasi* Coquand, p. 159.
1969 *Scaphites hippocrepis* (DeKay); Cobban, p. 17.
1969 *Scaphites hippocrepis* (DeKay) I; Cobban, p. 18, pl. 1, figs. 32–47; pl. 2, figs. 1–17; pl. 4, figs. 13–34; pl. 5, figs. 28–32; text-figs. 5–7, 9, 10, 17 (with full synonymy).
1969 *Scaphites hippocrepis* (DeKay) II; Cobban, p. 19, pl. 2, figs. 18–37; pl. 5, figs. 5–18, 24–27, 33–35; text-fig. 18 (with full synonymy).
1969 *Scaphites hippocrepis* (DeKay) III; Cobban, p. 21, pl. 3, figs. 1–25; pl. 4, figs. 35–49; pl. 5, figs. 36–40; text-figs. 2, 4, 10, 11 (with full synonymy).
1972 *Scaphites genyi* Lapeyre and Thomel,-p. 45, pls. 1 and 2.
1975 *Scaphites hippocrepis* (DeKay, 1827); Schmid and Ernst, p. 322, pl. 1, figs. 1 and 2.
1976 *Scaphites hippocrepis* (DeKay); Kennedy and Cobban, pl. 1, figs. 5 and 6.

Type. The holotype, by monotypy, is the original of DeKay 1828, pl. 5, fig. 5, a fragment from the Campanian of New Jersey. It is lost (Reeside 1962, p. 126). Reeside refers to the holotype of *Scaphites cuvieri* as serving as 'a standard of reference'. In the interests of nomenclatural stability I here designate this specimen neotype of the species. It is no. 19483 in the Collections of the Academy of Natural Sciences in Philadelphia, is from the Deep Cut, Delaware, and has been refigured by several authors (Whitfield 1892, pl. 44, fig. 8; Richards 1953, fig. 224; Reeside 1962, pl. 71, figs. 1–5).

Material. MAFI unregistered *ex* Coquand Collection, from Lavalette (Charente), the holotype of *S. nanclasi* Coquand, 1859. SP *ex* Arnaud Collection, numerous unregistered specimens: 15 are from P^1 or P^{1-2} at Livernant (Charente); 2 from P^1–P^2 at Moulin Batraud (Charente) (including the original of de Grossouvre 1894, pl. 32, fig. 3); route de Livernant à Montmoreau (Charente); 3 from P^2, La Caillerie, route de Girat (Charente) (= de Grossouvre 1894, pl. 37, fig. 3, who gives the locality as Mensignac (Dordogne)); 1 from P^{1-2}, Montmoreau (Charente); 4 from St-Hilaire-de-Montmoreau (Charente) (including the original of de Grossouvre 1894, pl. 32, fig. 2); 1 from P^1, La Massoulie (Dordogne); 9 from P^1–P^2, tranchée de Vaure, Mensignac (Dordogne); 1 from P^1, Puy de Fourches (Dordogne).

Description. Highly variable. Microconchs 18·5–31·0 mm long. Spire involute with compressed rounded whorls ornamented by strong, distant primary ribs which increase by bi- and trifurcation on the outer flank with additional intercalatories. Small ventrolateral tubercles present on later part of spire or not. Flank ribbing generally weakens or disappears on the narrow shaft which has a concave umbilical wall. Strong ventrolateral clavi extend to the end of the body-chamber. Small specimens have weak or no umbilical tubercles; large ones have conical to clavate umbilical tubercles throughout. Venter smooth or with feeble ribs in small specimens, effaced on first part or not in others; where present varying from fine to coarse, with pairs of distant ribs linking clavi, with or without single intercalatories. Aperture constricted.

Macroconchs 30–60 mm long, stouter than microconchs, with umbilical bulge occluding umbilicus, swollen and broadened at juncture of hook and shaft. Ornament on spire as in microconch. Shaft

---

EXPLANATION OF PLATE 19

Figs. 1–34. *Scaphites* (*Scaphites*) *hippocrepis* (DeKay, 1828). All SP unregistered *ex* Arnaud Collection. 1–6, 9–10, 18, 19, 24–26, 31, 32, from Assise P^1, 11–13, 20–21, from Assise P^1–P^2, Livernant (Charente); 7 and 8, are from Assise P^1–P^2, Mensignac-Vaure (Dordogne); 14–17, from Assise P^1, Puy-des-Fourches (Dordogne); 22, from Assise P^1, La Massoulie (Dordogne); 23, from Assise P^2, La Caillerie, route de Livernant à Montmoreau (Charente); 27–30, from Assise P^1, Mensignac-Vaure (Dordogne); 33–34, from Assise P^2, Moulin Batraud (Charente).

Figs. 35–37. *S.* (*Scaphites*) sp. SP unregistered *ex* Arnaud Collection, from Assise P^2, Chavenat (Charente). All figs. ×1.

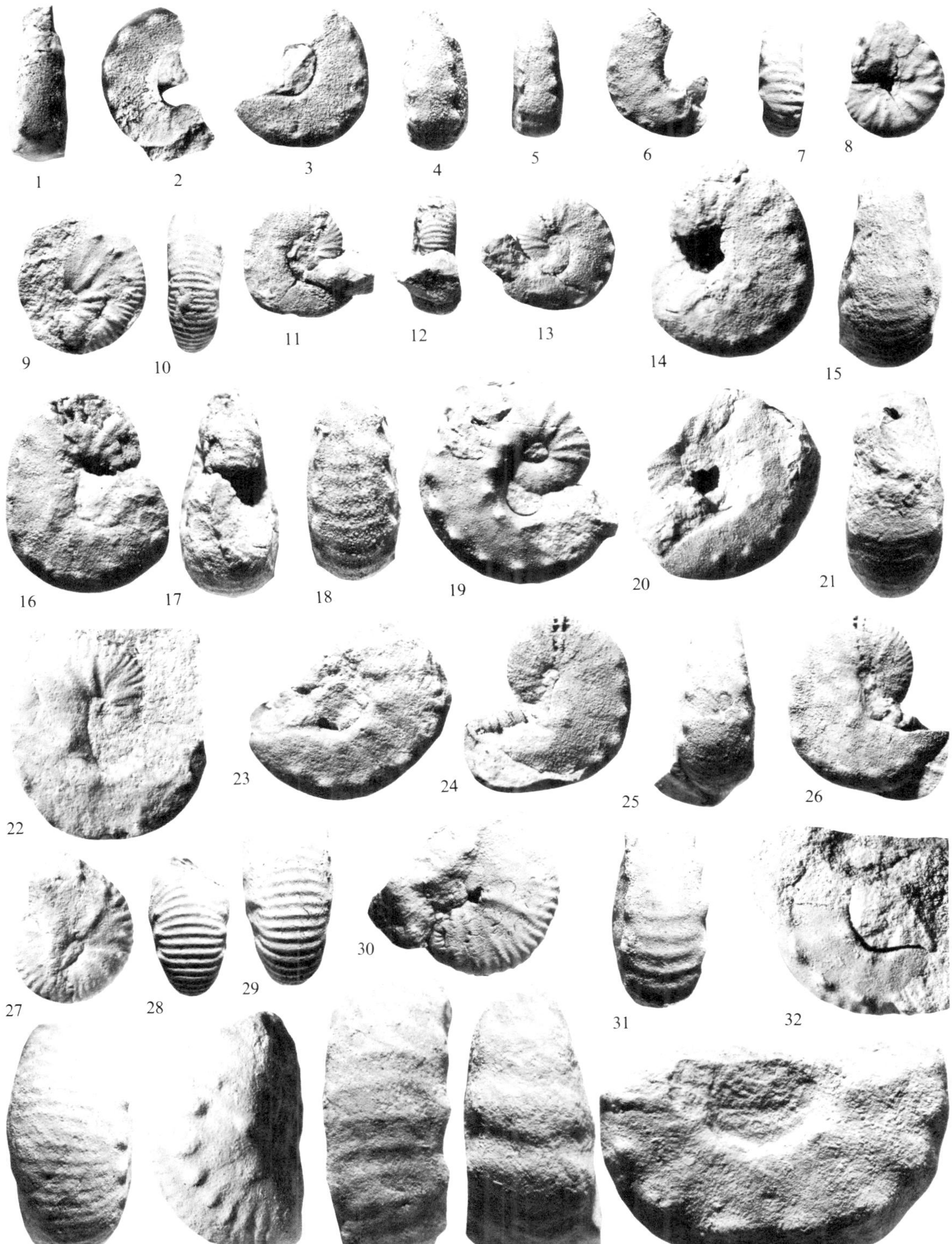

KENNEDY, *Scaphites* (*Scaphites*)

variable, with up to twelve conical to clavate ventral tubercles which reduce in size towards aperture. There are three to five umbilical tubercles, that at the juncture of hook and shaft much enlarged. Flank ornament varies from effaced to low, coarse, blunt ribs to well-differentiated ribs. A row of feeble or incipient lateral tubercles may appear at the end of the shaft and on the hook. Venter bears coarse to very fine ribs looping between tubercles or intercalating; commonly effaced on the middle part of the shaft. Extremes range from specimens with a single rib only to those with up to four linking tubercles. Aperture constricted.

Sutures not seen.

Discussion. This highly variable and strongly dimorphic species is relatively common in old collections and is of great importance for inter-regional correlation in the Lower Campanian. Cobban (1969) analysed the evolution of the US Western Interior representatives of this group in great detail. He recognized two species, *S. leei* Reeside, 1927 and its descendant *S. hippocrepis* DeKay, 1828, dividing each into three successive subspecies which he did not name but simple labelled forms I, II, and III. He found all forms to be strongly dimorphic with macroconchs and microconchs varying in size by factors of 1 to 4 in each form, the two overlapping in size but distinguished by the slender body-chamber, open umbilicus, and concave umbilical wall of microconchs and stout body-chamber, occluded umbilicus and convex umbilical wall of macroconchs. *S. leei* I has a coarsely ribbed body-chamber and ventrolateral tubercles on both phragmocone and body-chamber. *S. leei* II lacks tubercles on the phragmocone. *S. leei* III is more densely ribbed than *S. leei* II and macroconchs have fewer umbilical tubercles. *S. hippocrepis* I is more densely ribbed than its ancestor *S. leei* III and has ventrolateral tubercles that are weak and bullate rather than conical. There are no tubercles on the phragmocone, but microconchs have 2–4 umbilical and 5–10 ventrolateral tubercles; in macroconchs, the figures are 1–2 and 5–8. There are on average 25 ribs on the last part of the body-chamber in microconchs, and 22 in macroconchs.

S. hippocrepis II has well-defined ventrolateral tubercles and rather dense ribbing, which is weak or absent in small adults. This and ventrolateral tubercles that are rounded rather than bullate distinguishes it from *S. hippocrepis* I. *S. hippocrepis* III has numerous ventral ribs on the body chamber and may have weak node-like ribs on the middle of the flank between the umbilical and ventrolateral tubercles. In some individuals the ventrolateral tubercles extend on to the phragmocone. Microconchs have flattened flanks. There are 2–4 umbilical and 5–10 ventrolateral tubercles; the latter may be clavate. Macroconchs have 1–3 bullate umbilical tubercles and 5–10 small ventrolateral tubercles. Ventral ribs are evenly spaced and number 22–45 in microconchs and 22–52 in macroconchs.

The large (52 mm) macroconch from P^{1-2} at Montmoreau figured by de Grossouvre (1894, pl. 37, fig. 3; Pl. 20, figs. 3 and 4) and a second, larger body-chamber from P^2 on the route de Girat (Pl. 20, figs. 10 and 11) are very finely ribbed and show a distinct lateral tubercle, closely matching American specimens (e.g. Cobban 1969, pl. 3, figs. 22–25) of *S. hippocrepis* III. Other macroconchs lack such a lateral node, and are coarsely ribbed, but have tubercles on the spire (e.g. de Grossouvre 1894, pl. 32, figs. 2 and 3; Pl. 20, figs. 12–14). Cobban (1969, p. 24) took this to indicate them to be variants of *S. hippocrepis* III; such individuals are from P^2 at Moulin Batraud, St Hilaire-de-Montmoreau and Mensignac-Vaure.

EXPLANATION OF PLATE 20

Figs. 1–6, 10–19. *Scaphites* (*Scaphites*) *hippocrepis* (DeKay, 1828). 1–6, 10, 11, 15–19 are all SP unregistered, *ex* Arnaud Collection; 1 and 2, from Assise P^1, Livernant (Charente); 3 and 4, from Assise P^2, La Caillerie, route de Girat (Dordogne); 5, 6, 15–17, from Assise P^2, St Hilaire [-de-Montmoreau] (Charente); 18 and 19, from Assise P, Mensignac (Dordogne). 12–14, EMP unregistered from an unknown horizon and locality in Aquitaine, but illustrated by de Grossouvre 1894, pl. 35, fig. 6.

Figs. 7–9. *Nostoceras* (*Nostoceras*) *hyatti* Stephenson, 1941. SP unregistered *ex* Arnaud Collection, from Assise R^2 of Arnaud, Maurens (Dordogne).

All figs. ×1.

KENNEDY, *Scaphites* (*Scaphites*), *Nostoceras* (*Nostoceras*)

The Aquitaine assemblage thus clearly includes specimens with many features in common with some American individuals of *S. hippocrepis* III; these come from P^2. The same unit yields specimens with different combinations of ornament. I am particularly struck by the similarities between specimens from the Atlantic Seaboard (e.g. Reeside 1962, pl. 70, figs. 11 and 12; pl. 71, figs. 1–7; Cobban 1969, pl. 5, figs. 36–40) and those from Texas (e.g. Young 1963, p. 49, pl. 80, figs. 3 and 4).

Also referred to *S. hippocrepis* here is the largest assemblage, 15 specimens from Assise P^1 *sensu lato* at Livernant (Pl. 13, fig. 5; Pl. 19, figs. 1–6, 12–21, 23–26, 31, 32; Pl. 20, figs. 1 and 2). There are 9 microconchs, 3 macroconchs, and 3 juveniles. The microconchs vary from 18·5 to 32·0 mm in length. Large individuals have coarse primary ribs on the spire with 2–3 secondaries and may or may not have ventrolateral tubercles on the last part of the phragmocone. Ventrolateral clavi extend the whole length of the body-chamber in all individuals and may round slightly on the final hook. There are up to 10 clavate umbilical tubercles, and there may be faint, low, radial flank ribs. Ventral ribbing varies greatly from near-effacement in the smallest specimen to fine or coarse, with as few as one or two ribs linking the clavi. Some of the juvenile phragmocones of ? macroconchs have well-differentiated ventrolateral tubercles; others lack them. The one complete macroconch is only 27 mm long; the other two are body-chambers of specimens an estimated 45 mm long when complete. There are 2–3 conical umbilical tubercles, that at the juncture of shaft and hook being the largest, and an estimated 10 ventral tubercles that vary from conical to feebly clavate. There are no flank ribs but quite coarse ventral ribs, looping between ventral clavi and intercalating.

This assemblage shows characters of both Cobban's *S. hippocrepis* II and III; combining a lack of an incipient lateral node with, in some cases, many umbilical tubercles, consistently clavate outer ventrolateral tubercles in the microconchs and some macroconchs and tubercles on the spire in some individuals. De Grossouvre (1894, pl. 31, figs. 3, 4, 6) figured 3 microconchs from this locality as *S. aquisgranensis* Schlüter, 1872*a*; Cobban (1969, p. 24) regarded them as closest to rare forms of *S. hippocrepis* II (1969, pl. 5, figs. 19–23).

The Aquitaine succession thus records an evolutionary sequence that matches in certain respects the *S. hippocrepis* II → III transition documented by Cobban, although individual specimens from Europe do not find a match in the US Western Interior material. It is quite impossible from the present material to determine whether the name *aquisgranensis* can be usefully applied to part of the European lineage. I was unable to find any of Schlüter's syntypes of this species in Bonn, and there is no recent work on topotypes from the Vaals Formation that is their source. This unit certainly yields macroconchs that look inseparable from the neotype of *hippocrepis* (e.g. Holzapfel 1888, pl. 5, fig. 3; Van der Weijden 1943, pl. 13, fig. 8).

S. hippocrepis most closely resembles *S. haugi* de Grossouvre, 1894 among other species found in the Aquitaine Basin. Only macroconchs are known of this latter species and they have a distinctive trapezoidal costal section to the spire, with distant primary ribs which arise alternately at and just outside the umbilical shoulder and all bear strong ventral tubercles (Pl. 21, figs. 12–16). The clavi are very coarse on the body-chamber, while the flank ornament is low broad ribs with few massive umbilical nodes.

S. hippocrepis differs markedly from most other European *Scaphites*. Müller and Wollemann (1906, pl. 9, figs. 4 and 5; pl. 10, fig. 4) figured, as *S. binodosus*, microconchs (pl. 9, figs. 4 and 5) and a

---

EXPLANATION OF PLATE 21

Figs. 1–11. *Scaphites* (*Scaphites*) *hippocrepis* (DeKay, 1828). All SP unregistered *ex* Arnaud Collection. 1–3, from Assise P^1, Mensignac-Vaure (Dordogne); 4, 7, 9, 10, from Assise P^1–P^2 at the same locality; 8 and 11, from Assise P^2, St Hilaire [-de-Montmoreau] (Charente).

Figs. 12–16. *S.* (*S.*) *haugi* (de Grossouvre, 1894). 12–14, the holotype, 16, a topotype, SP unregistered *ex* Arnaud Collection, from Assise P^3, Tauillard (Charente); 15, no. 83421 in the J. P. Platel Collection from that author's Campanian 3, near Périgueux (Dordogne).

All figs. × 1.

KENNEDY, *Scaphites* (*Scaphites*)

TEXT-FIG. 40. A–H, *Scaphites (Scaphites) fischeri* Riedel, 1931. A, B, F–H, macroconchs; C–E, microconchs; note aptychus in H. All GPIB unregistered, Lower Campanian of Broitzem, Braunschweig, German Federal Republic, × 1.

macroconch (pl. 10, fig. 4) that I refer to a single species, selecting the name *S. fischeri* Riedel, 1931 (p. 704, pl. 79, figs. 5 and 6) (of which the lectotype, here designated is the original of Riedel 1931, pl. 79, fig. 6) for this form. The Collections of the Museum für Naturkünde, Berlin and GPIB contains a range of specimens from Braunschweig that I take to be the micro- and macroconch of this species: a selection are shown in text-fig. 40. Microconchs are up to 60 mm long. All have bullate to clavate umbilical tubercles extending from the end of the phragmocone to the end of the body-chamber (text-fig. 40C–E).

Ventrolateral tubercles occur on all of the outer whorl of the spire, strengthen into clavi on the shaft, and only efface slightly towards the adult aperture. The ribs, strong on the spire, weaken and may disappear on the body-chamber. Macroconchs (text-fig. 40A, B, F, G) are 50–80 mm long. Ventral tubercles occur on the outer whorl of the spire, strengthen into clavi on the body-chamber, where there are persistent umbilical tubercles and weak or no ribs. Some of these specimens appear to be giant counterparts of some Aquitaine specimens (e.g. Pl. 19, figs. 18 and 19), but French examples have a spire with distant primaries that generally branch on the outer flank or ventrolateral shoulder, no umbilical and variably developed ventrolateral tubercles. This interesting German assemblage thus includes specimens with both *hippocrepis* and *binodosus*-like characters.

Occurrence. Lower Campanian, Arnaud's specimens are labelled Assise P^{1-2} in Aquitaine, although de Grossouvre (1901. p. 378) states that the species occurs only in the upper part of P^1, his P^{1a}, later (1901, p. 801) regarding this as a local circumstance only. Contes (Alpes-Maritimes) in France, Hampshire in England (Gale 1980, p. 138), where it occurs in the upper part of the *Gonioteuthis quadrata* Zone; there are numerous specimens from the *quadrata* Zone Vaals Formation in Belgium, the Netherlands, and German Federal Republic as well as its correlatives elsewhere in the latter country; the careful records of Schmid and Ernst (1975, fig. 2) show it ranging to the top of their Lower Campanian in Westphalia. It may also occur in Sweden (Moberg 1855, p. 26, pl. 3, fig. 2). The species is widely distributed in the United States, from the Merchantville Formation in New Jersey and Delaware, Blufftown Formation in Alabama and Georgia, Austin Chalk in Texas, precisely localized in the Gober Tongue of Lamar County. In the US Western Interior it occurs in Montana, Wyoming, South Dakota, Colorado, Utah, and New Mexico, ranging through Zones of *S. hippocrepis* I, II, and III in the Lower Campanian, the base of the stage being taken below the underlying Zone of *S. leei* III.

Scaphites (*Scaphites*) sp.

Plate 19, figs. 35–37

Material. SP unregistered *ex* Arnaud Collection, from Assise P^2 at Chavenat (Charente).

Description. The specimen is a composite internal mould of the final hook of a body-chamber 50 mm wide. Whorl section depressed (whorl breadth to height ratio 1·36). Umbilical wall is concave, umbilical shoulder narrowly rounded. Flanks flat, subparallel in costal section, venter broadly rounded. There are four strong umbilical bullae. These give rise to pairs of low, broad, coarse ribs that terminate in strong, prominent ventrolateral clavi that weaken markedly towards the aperture. They are linked over the venter by broad coarse folds at the end of the shaft (Pl. 19, fig. 35); on the hook there are pairs of strong narrow convex ribs looping between clavi with occasional intercalatories (Pl. 19, fig. 37).

Discussion. The narrowness of the whorl and concave umbilical wall show this to be a giant microconch. It most closely resembles the microconchs of early forms of *Scaphites hippocrepis* (compare Pl. 19, fig. 19 and Pl. 19, fig. 37), but if a basically similar shell form is assumed it must have been nearly 60 mm long versus 18·5–31·0 mm. With such limited material it is difficult to speculate further: it may well be a new species.

Occurrence. As for material.

Scaphites (*Scaphites*) *gibbus* Schlüter, 1872*a*

Plate 22, figs. 1, 2, 5–8; text-fig. 41

1872*a* *Scaphites gibbus* Schlüter, p. 87, pl. 26, figs. 6–9.
1889 *Scaphites gibbus* Schlüter; Griepenkerl, p. 404 (102).
1894 *Scaphites gibbus* Schlüter; de Grossouvre, p. 251, pl. 32, fig. 10.
1911 ?*Acanthoscaphites gibbus* Schlüter; Nowak, p. 565.
1916 *Hoploscaphites gibbus* Schlüter; Nowak, p. 66.
1925 *Discoscaphites gibbus* Schlüter; Diener, p. 211.
1928 *Discoscaphites gibbus* (Schlüter); Reeside, p. 30.
1951 *Discoscaphites gibbus* (Schlüter); Mikhailov, p. 94, pl. 18, figs. 86 and 87.
1959 *Discoscaphites gibbus* (Schlüter); Naidin and Shimanskij, p. 197, pl. 6, figs. 9–11.

1964 *Hoploscaphites gibbus* (Schlüter); Giers, p. 283, pl. 8, fig. 3; text-fig. 10.
1974 *Trachyscaphites* (?) *gibbus* (Schlüter, 1872); Naidin, p. 172, pl. 58, figs. 2–4.
1975 *Scaphites gibbus* Schlüter, 1872; Schmid and Ernst, p. 327, pl. 1, figs. 3–6.
1976 ?*Trachyscaphites gibbus* (Schlüter); Atabekian and Khakhimov, p. 71, pl. 8, figs. 4–6 (with additional synonymy).
1980 *Trachyscaphites* (?) *gibbus* (Schlüter); Błaszkiewicz, p. 32, pl. 13, figs. 6 and 8.

Types. Schlüter refers to thirty specimens and figured two (1872*a*, pl. 26, figs. 6–9). Atabekian and Khakhimov (1976, p. 71) refer to the original of figs. 7–9 as the holotype, but I take this not to be a valid lectotype designation. Błaszkiewicz (1980, p. 32) has designated the same specimen lectotype. It is GPIB 63, a macroconch from the Campanian of Baumberg between Coesfeld and Billerbeck, Westphalia, German Federal Republic (text-fig. 41). The figured paralectotype (Schlüter 1872*a*, pl. 26, fig. 6) seems to be a very small macroconch, and is an unregistered specimen in the same collection and from Darup, Westphalia.

Material. SP unregistered *ex* Arnaud Collection: four specimens from Assise P^3 at St-Médard-de-Barbezieux (Charente).

Description. All specimens are crushed composite internal moulds. One is a microconch 63 mm long, the remainder macroconchs 70–81 mm long. Spire of microconch very involute with a tiny umbilicus. Fine feebly flexuous prorsiradiate primary ribs arise, either singly or in pairs, from umbilical shoulder and increase by branching or intercalation so that rib density is uniform over flanks. Occasional feeble tubercles develop at point of branching, while there are strong ventral clavi on all of exposed outer whorl linked over venter by 2–3 ribs and separated by 1–2 ribs of comparable strength. Umbilicus not occluded by shaft; flank ribbing effaced, with 2 blunt umbilical nodes, that at the curvature of the final hook expanded into a prominent umbilical swelling. Strong ventral clavi persist to the beginning of curved portion, linked over venter by 2–3 riblets. Coarse flank ribs rejuvenate on final hook, apparently without tubercles, although preservation is defective. Spire of macroconch with coarser primary ribs, strengthened in some cases into feeble umbilical bullae, increasing by branching low on flank. Bullate to clavate ventrolateral tubercles are connected over the venter by up to 3 ribs and separated by up to 5 primary, secondary, and intercalated ribs. Shaft partially occludes umbilicus. Flank ribs efface. A massive umbilical bulge develops at beginning of final hook. Lateral tuberculation variable, with an inner lateral and 2 outer lateral rows of bullate to conical nodes that strengthen as size increases. Ventrolateral clavi strong, linked over venter by up to 5 ribs with 2–3 equal ribs between. Lateral tubercles strengthen while ventral clavi reduce and change from elongate to conical on final hook, the latter linked over venter by coarsening ribs so that, ultimately, a single rib links tubercles with a single intercalatory. Flank ornament at aperture may include up to 5 rows of small tubercles; actual aperture constricted, with dorsal projection.

Sutures not seen.

Discussion. Schlüter's figure of the lectotype shows ventral tubercles at the end of the phragmocone only and a maximum of three rows of lateral tubercles, while the figured macroconch paralectotype lacks ventrolateral clavi on the spire. I take these differences as no more than intraspecific variation as is supported by the variable ornament of material from the Lehrte syncline, Westphalia (Schmid and Ernst 1975, pl. 1, figs. 3–6).

This is a highly distinctive species. The coiling and tuberculation immediately separate it from all other species occurring in western Europe. It most closely resembles *Scaphites haugi* de Grossouvre, 1894 (p. 244, pl. 31, fig. 5). This has a spire with coarse single flank ribs that terminate in conical ventrolateral tubercles linked over the venter by convex secondaries and a shaft with coarse umbilical and ventrolateral tubercles only.

EXPLANATION OF PLATE 22

Figs. 1, 2, 5–8. *Scaphites* (*Scaphites*) *gibbus* Schlüter, 1872*a*. SP unregistered *ex* Arnaud Collection, from Assise P^3, St Médard-de-Barbezieux (Charente).

Fig. 3. *Nostoceras* (*Bostrychoceras*) *polyplocum* (Roemer, 1841). SP unregistered *ex* Arnaud Collection, from Assise Q, Rioux Martin (Charente).

Fig. 4. *Trachyscaphites spiniger* (Schlüter, 1872*a*). SP unregistered *ex* Arnaud Collection, from St Médard [-de-Barbezieux] (Charente).

All figs. × 1.

PLATE 22

KENNEDY, *Scaphites* (*Scaphites*), *Nostoceras* (*Bostrychoceras*), *Trachyscaphites*

TEXT-FIG. 41. *Scaphites (Scaphites) gibbus* (Schlüter, 1872*a*). GPIB 63, the lectotype, a macroconch, the original of Schlüter 1872*a*, pl. 26, figs. 7–9, × 1.

As will be seen from the synonymy, there is no consensus on the generic assignation of this species. It precedes and overlaps with *Trachyscaphites spiniger* (Schlüter, 1872*a*) in Germany (Schmid and Ernst 1975, fig. 2) while it shows an intermediate state of lateral tuberculation between that of typical *Scaphites* and *Trachyscaphites*. Whereas the ventrolateral clavi are alternate in *Trachyscaphites* and some possess a siphonal tubercle, the clavi are opposed in the present species and no siphonal tubercle develops. Furthermore the species shows marked differentiation of ribbing, lack of which Cobban and Scott stressed (1964, p. E6–7) when erecting their new genus, and shows the hump-backed coiling of earlier *Scaphites* with a short shaft and aperture close to spire, rather than the straight shaft and widely separated aperture and spire of *Trachyscaphites*. It is obviously far-removed from the recoiled hypermorphic giant *Acanthoscaphites* Nowak, 1911 (p. 565; type species *S. tridens* Kner, 1848, p. 10,

---

EXPLANATION OF PLATE 23

Figs. 1 and 7. *Baculites* sp. 1. SP unregistered *ex* Arnaud Collection. 1 is from Assise P^2, St Hilaire [-de-Montmoreau] (Charente); 7 is from Assise P^2, La Caillerie (Charente).

Figs. 2 and 3. *Scaphites (Scaphites) hippocrepis* (DeKay, 1828). MAFI unregistered *ex* Coquand Collection, the holotype of *S. nanclasi* Coquand, 1859, from Lavalette (Charente).

Figs. 4–6, 8–10, 14, 15, 19, 22, 23, 27. *Baculites* sp. 7. SP unregistered *ex* Arnaud Collection, from the Silex de Mussidan of Mussidan (Dordogne).

Figs. 11–13, 16–18, 20, 21. *Baculites* sp. 3. SP unregistered *ex* Arnaud Collection. 11–13, 20, 21 are from Assise Q at Issac (Dordogne); 16–18, from the same horizon at Rioux Martin (Dordogne).

Figs. 24–26. *Baculites* sp. 6. OUM KZ13863, from the lumachelle facies, Assise R, roadside ditches between La Petite Métairie and railway crossing by Chez Vardon, 2·2 km south of Baignes-Ste-Radegonde (Charente-Maritime).

Figs. 28 and 29. *Baculites* sp. 2. SP unregistered *ex* Arnaud Collection, from Assise P^2, Montmoreau (Charente).

All figs. × 1.

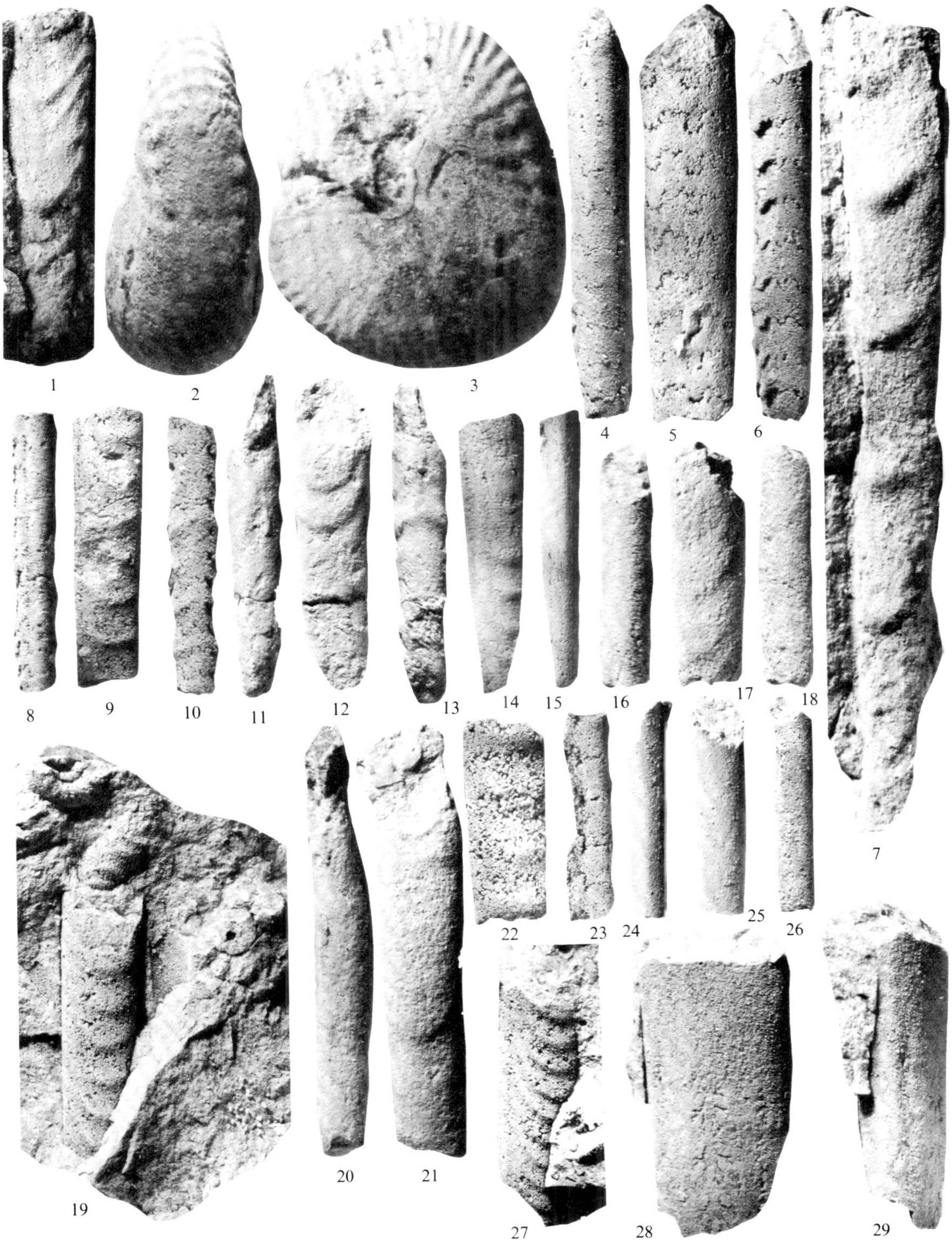

KENNEDY, *Baculites*, *Scaphites* (*Scaphites*)

pl. 2, fig. 1, by the subsequent designation of Diener 1925, p. 205). *Discoscaphites* Meek, 1871 (type species by original designation *Ammonites conradi* Morton, 1834, p. 39) has, in the type and closely allied American species at least, a quite different coiling pattern with a straight shaft, no umbilical bulge in macroconchs, up to nine rows of tubercles on each side, the number decreasing on the body-chamber. I strongly suspect that the various Old World species referred to this genus are simply multituberculate derivatives of other stocks.

Occurrence. Apart from the present records, de Grossouvre (1894, p. 252) noted a specimen from Angoumé (Landes). The species is generally regarded as lower Upper Campanian in north Germany, Poland, Donbass, Transcaspia, and Ukraine, USSR. The precise records of Schmid and Ernst (1975, fig. 2) show it spanning the Lower–Upper Campanian boundary in Lower Saxony.

Genus TRACHYSCAPHITES Cobban and Scott, 1964, p. E7

Type species. Trachyscaphites redbirdensis Cobban and Scott, 1964, p. E7, pl. 1, figs. 1–7; text-fig. 3, by original designation.

Trachyscaphites spiniger (Schlüter, 1872*a*)

Plate 22, fig. 4; text-fig. 42

1841 *Scaphites pulcherrimus* Roemer, p. 91 (*pars*), pl. 14, fig. 4.
1872*a* *Scaphites spiniger* Schlüter, p. 82, pl. 25, figs. 1–7.
1964 *Trachyscaphites spiniger* (Schlüter); Cobban and Scott, p. E9 (with full synonymy).
1974 *Trachyscaphites spiniger* (Schlüter); Naidin, p. 171, pl. 59, fig. 2.
1975 *Scaphites spiniger* Schlüter; Schmid and Ernst, p. 330, pl. 2, figs. 1–4; pl. 3, fig. 3.
1976 *Trachyscaphites spiniger spiniger* (Schlüter); Atabekian and Khakhimov, p. 66, pl. 9, figs. 1–4; pl. 12, figs. 2–4.
1980 *Trachyscaphites spiniger spiniger* (Schlüter); Błaszkiewicz, p. 30, pl. 13, figs. 1–3, 5, 7.
(Synonymy of nominate subspecies only.)

Types. The lectotype, by the subsequent designation of Błaszkiewicz 1980, p. 31 is the original of Schlüter 1872*a*, pl. 25, figs. 1–3, a macroconch from the Upper Campanian of Darup, Westphalia, German Federal Republic, an unregistered specimen in the GPIB Collections. Paralectotypes are GPIB 61*a*, the original of Schlüter 1872*a*, pl. 25, fig. 4, from the Hügelgruppe of Haldem, an adult microconch (text-fig. 42A, B), and GPIB 61*b*, the original of Schlüter 1872*a*, pl. 25, fig. 6, from the same horizon and locality, showing the aptychus (text-fig. 42F).

Material. SP unregistered *ex* Arnaud Collection, from Assise P^3, 'St Medard' [de-Barbezieux] (Charente).

Description. The specimen is a composite mould of part of the final hook. It retains traces of an outer ventrolateral row of strong clavi, with a slightly weaker, clavate inner ventrolateral row, a weaker, feebly clavate outer lateral, weak, conical mid-lateral and conical to feebly bullate umbilical tubercles, all borne on a low, broad rib.

Discussion. The fragment shows the tuberculation typical of *Scaphites spiniger*, although lacking, due perhaps to poor preservation, the delicate flank ribs so prominent in the German material (text-fig. 42). It differs from *T. pulcherrimus* (Roemer, 1841) the only other *Trachyscaphites* found in Aquitaine, by its four, rather than five rows of tubercles on each side. *T. spiniger* has been divided into several subspecies. *T. spiniger porchi* Adkins, 1929 (p. 205, pl. 5, figs. 1–3; see Cobban and Scott 1964, p. E10, pl. 2, figs. 1–23; pl. 3, figs. 1–11; text-fig. 4) from the Upper Campanian of Texas and the US Western Interior differs from the nominate subspecies in having fewer tubercles on the body-chamber, stronger

TEXT-FIG. 42. A–F, *Trachyscaphites spiniger* (Schlüter, 1872*a*). A, B, a paralectotype, a microconch, GPIB 61*a*, the original of Schlüter 1872*a*, pl. 25, fig. 4; C, MNB unregistered, with aptychus; D, MNB unregistered, an adult macroconch; E, MNB unregistered, an uncrushed microconch; F, GPIB 61, the original of Schlüter 1872*a*, pl. 25, figs. 5 and 6, a paralectotype with aptychus. All specimens from the Upper Campanian of Haldem, Westphalia, German Federal Republic, × 1.

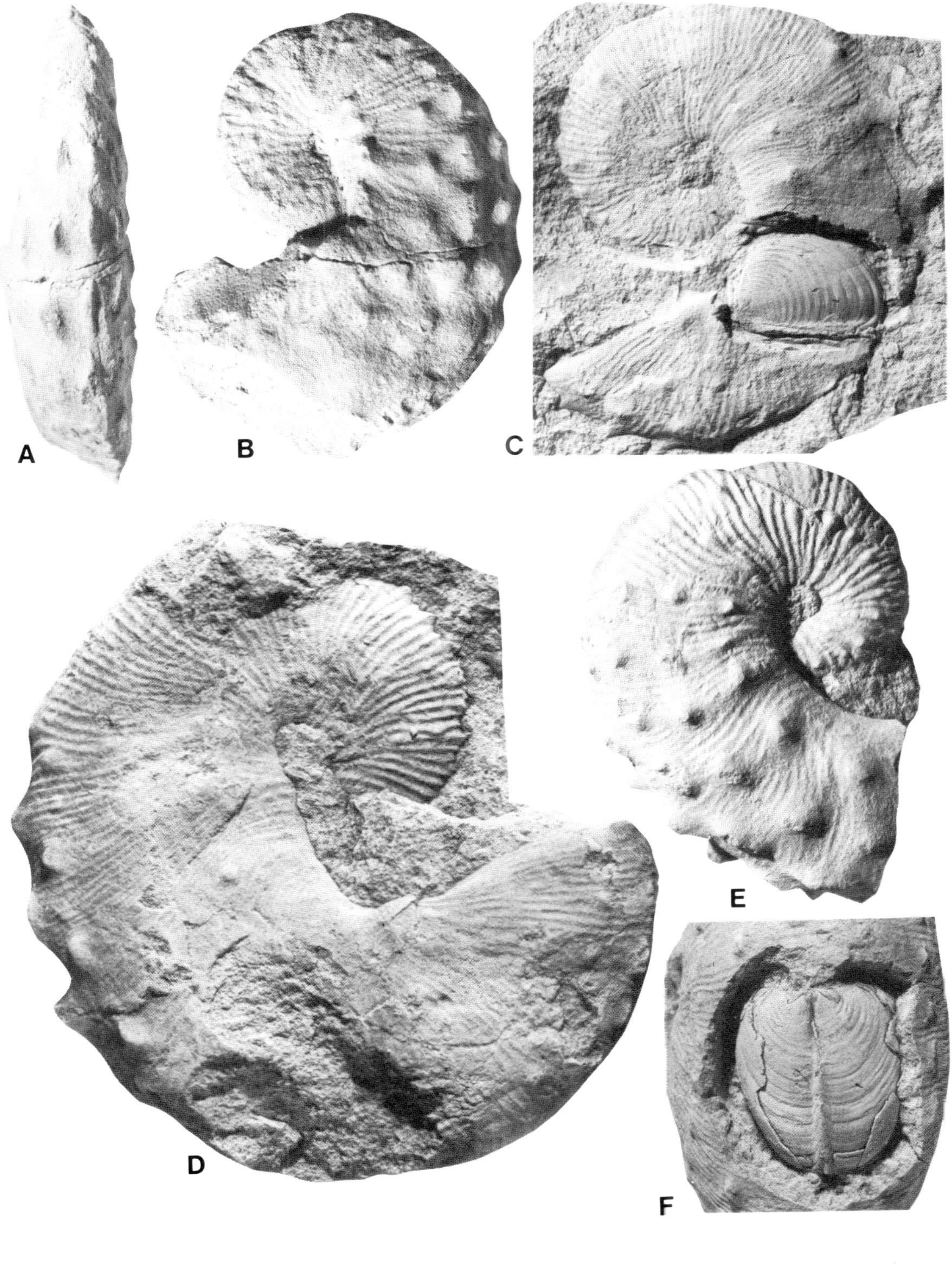
A
B
C
D
E
F

tuberculation, and weaker ribbing. *T. spiniger posterior* Błaszkiewicz, 1980 (p. 31, pl. 13, fig. 4; pl. 14, figs. 1–7; pl. 15, figs. 2 and 3; pl. 30, fig. 2) from the Upper Campanian of the Middle Vistula Valley, Poland was separated from the nominate subspecies by its author because of the smaller number of ribs linking the rows of tubercles on the spire, presence of umbilical tubercles at small diameters as well as the form of coiling. *T. spiniger levantinensis* Lewy, 1969 (p. 132, pl. 4, fig. 1) was separated from the nominate species by its author because of the absence of ribs on the body-chamber. It is from the Upper Campanian of Jerusalem, Israel.

It is impossible to evaluate these subspecies until the abundant and variable German material is revised. The present specimen shows the tuberculation of *spiniger sensu stricto*, but the lack of ribs recalls both *porchi* and *levantinensis*.

Occurrence. The nominate subspecies is known from the Upper Campanian of the German Federal Republic, the Netherlands, Sweden, Poland, and the USSR (European Russia, Soviet Armenian, Kopet Dag). I am uncertain of the affinities of the specimen from the Craie de Spiennes in the Mons Basin, Belgium, figured by de Grossouvre (1908*a*, p. 38, pl. 10, fig. 6; text-fig. 13) and have not seen the specimens cited but not figured by that author from the Craie de Meudon, Hauts-de-Seine, France. Cobban and Scott (1964, pp. E5–6) have already commented on the problem of precise range of this species. The most recent work shows it as a lower Upper Campanian *marroti* Zone species in Westphalia (Schmid and Ernst 1975, fig. 2) and the Kopet Dag (Atabekian 1979, table 1). In Poland it is similarly limited to the lower Upper Campanian *phaleratum* Zone of Błaszkiewicz (1980, table 1), with subspecies *posterior* occurring in the succeeding *polyplocum* Zone. Subspecies *levantinensis* occurs in Israel, *porchi* in Texas, Montana, South Dakota, Wyoming, Utah, and Colorado, USA.

Trachyscaphites pulcherrimus (Roemer, 1841)

Plate 13, fig. 4; Plate 15, fig. 4; Plate 16, figs. 8, 9, 12–17; text-fig. 43

1841 *Scaphites pulcherrimus* Roemer, p. 91 (*pars*), *non* pl. 14, fig. 4.
1866 *Scaphites multinodosus* Hauer, p. 306, pl. 1, figs. 7 and 8.
1872*a* *Scaphites pulcherrimus* Roemer; Schlüter, p. 85, pl. 26, figs. 1–5.
1894 *Scaphites pulcherrimus* Roemer sp.; de Grossouvre, p. 250, pl. 32, figs. 6 and 9.
1915 *Scaphites spinosissimus* Frech, p. 564, text-fig. 12.
1916 *Acanthoscaphites pulcherrimus* (Roemer); Nowak, p. 63.
1925 *Scaphites (Acanthoscaphites) pulcherrimus* Roemer; Diener, p. 206.
1928 *Acanthoscaphites pulcherrimus* Roemer; Reeside, p. 33.
1951 *Acanthoscaphites pulcherrimus* (Roemer); Mikhailov, p. 96, pl. 18, figs. 83 and 84.
1959 *Acanthoscaphites pulcherrimus* (Roemer); Naidin and Shimanskij, p. 195, pl. 6, fig. 14.
1964 *Trachyscaphites pulcherrimus* (Roemer); Cobban and Scott, p. 1.
1968 *Scaphites pulcherrimus* Roemer; Arnold, p. 314, text-fig. 36.
1972 *Trachyscaphites pulcherrimus* (Roemer); Khakhimov, p. 160, pl. 1, fig. 3.
1973*b* *Trachyscaphites pulcherrimus* (Roemer); Cobban, p. 695, figs. 1*a–g*, *k*, *l*, 2*a–o*.
1974 *Trachyscaphites pulcherrimus* (Roemer, 1841); Naidin, p. 171, pl. 58, fig. 5.
1976 *Trachyscaphites pulcherrimus* (Roemer); Atabekian and Khakhimov, p. 68, pl. 8, fig. 3; pl. 12, fig. 1.
1980 *Trachyscaphites pulcherrimus* (Roemer, 1841); Błaszkiewicz, p. 33, pl. 15, figs. 1, 4–11.
1982 *Trachyscaphites pulcherrimus* Roemer; Martinez, p. 173, pl. 30, fig. 7.
1984 *Trachyscaphites pulcherrimus* (Roemer, 1841); Kennedy and Summesberger, p. 171, pl. 11, figs. 1 and 2; pl. 12, figs. 1–7, 10–22; pl. 13, figs. 2–6.

Types. Roemer based *pulcherrimus* on several specimens, which he described as having five rows of tubercles on each side. The figured one (1841, pl. 14, fig. 4) has only four and was renamed *Scaphites spiniger* by Schlüter (1872*a*, p. 86). Cobban (1973*b*, explanation of fig. 1*k*, *l*) referred to the original of Schlüter 1872*a*, pl. 26, figs. 1–3 as the holotype of *pulcherrimus* but it is not, because it was not one of Roemer's original specimens. Summesberger and I (1984) presumed the Roemer types to be in the Roemer–Pelizaeus Museum, Hildersheim, but this has proved not to be the case. Some of Roemer's 1840–1841 Cretaceous types were in Breslau and studied by Frech (1915) and a few still survive in the Geological Museum of Wroclaw University (letter from Dr J. Gorczyca-Skala, 27.11.1984), although none of the syntypes of the present species. In the interests of nomenclatural stability I here designate GBIB 62, the original of Schlüter 1872*a*, pl. 26, figs. 1–3, a macroconch from the Upper Campanian of Haldem, Westphalia, German Federal Republic, neotype of the species. It is a composite mould of a macroconch and is shown in text-fig. 43B.

TEXT-FIG. 43. A–G, *Trachyscaphites pulcherrimus* (Roemer, 1841). A, GPIB 62a, the original of Schlüter 1872*a*, pl. 26, figs. 4 and 5, a microconch; B, neotype, GPIB 62, the original of Schlüter 1872*a*, pl. 26, figs. 1–3, a macroconch; C–E, MNB unregistered, a complete microconch; F, G, MNB unregistered, a macroconch. All specimens from the Upper Campanian of Haldem, Westphalia, German Federal Republic, ×1.

Material. SP unregistered *ex* Arnaud Collection, 7 specimens: 3 are from Assise P^3 at Le Caillaud, the cliff section south of Talmont (Charente-Maritime); 1 is labelled QR, Courgeac-Beauvais and is the original of de Grossouvre 1894, pl. 32, fig. 6 (where it is said to be from Assise Q at Neuvic); 1 is labelled Assise Q, Neuvic (Dordogne) and is the original of de Grossouvre 1894, pl. 32, fig. 9; 1 is simply labelled 'Meschers' [-sur-Gironde] (Charente-Maritime); 1 is from Assise P^3, St Médard [de-Barbezieux] (Charente) and identified as *Heteroceras polyplocum* (Pl. 15, fig. 4) by Arnaud.

Description. All specimens are crushed composite moulds. Four are microconchs, with maximum length 63–67 mm; one is a fragmentary macroconch with an estimated length of 100 mm. Coiled phragmocone compressed; umbilicus small, deep, umbilical shoulder narrowly rounded, inner flank broadly rounded, outer flank flattened and convergent, ventrolateral shoulder faceted, venter flat in costal section, rounded intercostally. Elongate bullae perch on umbilical shoulder and give rise to one or two narrow flexuous prorsiradiate primary ribs which branch irregularly once or twice and increase by intercalation, so that rib spacing is constant across flanks. There are small conical inner and outer lateral and small feebly clavate inner and outer ventrolateral tubercles. The latter alternate on either side of the venter and are linked by irregular zig-zag ribs with a small siphonal tubercle. The straight part of the shaft bears a full complement of lateral and ventrolateral tubercles and shows ribs looping between the tubercles in some cases. The siphonal tubercle may decline or persist. The final hook shows a coarsening of ribs at the expense of tubercles and becomes much more markedly prorsiradiate; umbilical, inner and outer ventrolateral tubercles (the latter alternating when viewed in ventral profile) persist, but the laterals are ultimately lost.

The single macroconch is part of the final shaft and hook, with aperture and a feeble umbilical bulge. Sutures are only imperfectly exposed.

Discussion. This distinctive species is easily separated from all other European *Trachyscaphites* by the presence of five rows of lateral plus siphonal tubercles; the ribbing is also distinctive. *S. multinodosus* Hauer, 1866 (p. 306, pl. 1, figs. 7 and 8; see Kennedy and Summesberger 1984, pl. 12, figs. 1–3) is no more than a microconch *pulcherrimus* as is *S. spinosissimus* Frech, 1915 (p. 564, text-fig. 12).

There is some variation in the point of appearance and the persistence of siphonal tubercles in this species; with German material this is in part a reflection of preservation. There is also variation in adult size. Specimens from the Ultrahelvetic of the Gschliefgraben, Upper Austria (Kennedy and Summesberger 1984) have microconchs 30–50 mm long and macroconchs 64–68 mm long; material from Haldem, Westphalia, shows microconchs up to 55 mm and macroconchs up to 95 mm long, while the present material has microconchs 63–67 mm long.

Occurrence. *T. pulcherrimus* is confined to the Upper Campanian; the most reliable records show it occurring in the upper part of the substage, the *polyplocum* Zone and its correlatives, while in Poland it extends up into the succeeding *donezianum* Zone. The geographic range extends from the Aquitaine Basin to northern Spain (Lleida), the German Federal Republic, Austria (Gschliefgraben, Ultrahelvetic, not Gosau Basins), Poland, European Russia, Soviet Armenia, New Jersey and Texas, USA.

Acknowledgements. I thank J. M. Hancock and R. Parish for assistance in the field, C. W. Wright, W. A. Cobban, and U. Kaplan for valuable advice and criticism. M. Neumann and J. P. Platel provided invaluable information on the Aquitaine succession and showed me specimens from their own collections, as did Madame M. Séronie-Vivien. Many museum curators allowed me to study specimens in their care: I especially thank D. Pajaud, J. Sornay, J. Louail, G. Marie, P. Moreau, A. Prieur, R. Busnardo, J. P. Thieuloy, H. Jaeger, H. Remy, A. V. Dhondt, D. Phillips, H. G. Owen, M. K. Howarth, and C. J. Wood for their assistance. The technical assistance of the staff of the Geological Collections, University Museum and Department of Earth Sciences, Oxford is gratefully acknowledged, as is the financial support of the Royal Society and Natural Environment Research Council.

REFERENCES

ADKINS, W. S. 1928. Handbook of Texas Cretaceous fossils. *Bull. Univ. Tex.* **2838,** 385 pp., 37 pls.

—— 1929. Some Upper Cretaceous Taylor ammonites from Texas. Ibid. **2901,** 203–211, pls. 5 and 6.

—— 1933. Mesozoic Systems in Texas. *In* SELLARDS, E. H., ADKINS, W. S. and PLUMMER, F. B. (eds.). The Geology of Texas, vol. 1, stratigraphy. Ibid. **3232,** 240–518.

ANDERSON, F. M. 1958. Upper Cretaceous of the Pacific coast. *Mem. geol. Soc. Am.* **71,** xi + 378 pp., 75 pls.

ARCHIAC, A. D'. 1836. Mémoire sur la formation crétacée du Sud-Ouest de la France. *Mém. Soc. géol. Fr.* (1) **2,** 157–193, pls. 11–13.
—— 1843. Études sur la formation crétacée du Sud-Ouest de la France. *Annls Sci. géol. Paris*, **2,** 121–143.
—— 1851. *Histoire de progrès de la géologie*, **4,** 600 pp. Société géologique de France, Paris.
—— 1857. Observations à M. Coquand. *Bull. Soc. géol. Fr.* (2) **14,** 766–770.
—— 1859. Les Corbières. *Mém. Soc. géol. Fr.* (2) **6,** 209–446, pls. 1–6.
ARKELL, W. J. 1952. A monograph of English Bathonian ammonites. *Palaeontogr. Soc.* [*Monogr.*], Part 3, 73–102, pls. 9–11.
ARNAUD, H. 1862. Note sur la craie de la Dordogne. *Bull. Soc. géol. Fr.* (2) **19,** 465–500, pl. 11.
—— 1873. Profils géologiques des chemins de fer d'Orléans traversant la craie du Sud-Ouest. Ibid. (3) **1,** 405–408, pl. 9.
—— 1876. Profil géologique des falaises crétacées de la Gironde. Étude comparative sur le Dordonien. *Act. Soc. linn. Bordeaux*, **30,** 554–574.
—— 1877*a*. Profils géologiques des chemins de fer des Charentes, région crétacée. Études pratiques de la craie du Sud. Ouest, 2e partie. Ibid. **31,** 168–201.
—— 1877*b*. Mémoire sur le terrain crétacé du Sud-Ouest de la France. *Mém. Soc. géol. Fr.* (2) **10,** 110 pp., pls. 21–28 (1–8), tables.
—— 1877*c*. Profils géologiques des chemins de fer d'Orléans, région crétacée. *Act. Soc. linn. Bordeaux*, **31,** 251–286.
—— 1878*a*. Parallèlisme de la craie supérieur dans le Nord et le Sud-Ouest de la France. *Bull. Soc. géol. Fr.* (3) **6,** 205–211.
—— 1878*b*. Synchronisme de l'étage du Turonien dans le Sud-Ouest et le Midi de la France. Ibid. **6,** 233–242.
—— 1883*a*. Profils géologiques des chemins de fer de Siorac à Sarlat et de Périgueux à Ribérac. Études pratiques sur le craie de Sud-Ouest, 4e partie. *Acta. Soc. linn. Bordeaux*, **37,** 34–48, pls. 1–3.
—— 1883*b*. Parallèlisme entre la craie du Nord et celle du Sud-Ouest. *Bull. Soc. géol. Fr.* (3) **11,** 92.
—— 1887*a*. Aperçu générale sur la craie du Sud-Ouest. Ibid. **15,** 809–813, pl. 32.
—— 1887*b*. Compte-rendu de l'excursion à Saint-Cirq et de Saint-Cirq à Beaumont. Ibid. 841–845.
—— 1887*c*. Résumé générale des observations sur la craie du Sud-Ouest. Ibid. 884–911.
—— 1887*d*. Excursion à Montignac-sur-Vezère. Ibid. 914–918.
—— 1891. De la limite tracée par Coquand entre le Santonien et Campanien. Ibid. **19,** 665–668.
—— 1892. Profil géologique du chemin de fer d'Angoulême à Marmande, région crétacée. Études pratiques sur la craie de Sud-Ouest. 5e études. *Act. Soc. linn. Bordeaux*, **45,** 11–43, 1 pl.
—— 1896. Découverte de la *B. mucronata* près de Barbezieux (Charente). *Bull. Soc. géol. Fr.* (3) **24,** 85.
—— 1897. Divisions naturelles du Crétacé supérieur au-dessus du Santonien dans le Sud-Ouest et la région pyrénéenne. Ibid. **25,** 676–681.
ARNOLD, H. 1968. Das Obercampan des Stemweder Berges und seine Fauna. *Veröff. Uberseemus. Bremen*, **A3,** 272–342.
ATABEKIAN, A. A. 1979. Correlation of the Campanian stage in Kopet dag and western Europe. *Aspekte der Kreide Europas*, IUGS Series **A6,** 511–526.
—— and AKOPIAN, V. T. 1970. Late Cretaceous ammonites of the Armenian SSR (Pachydiscidae, Kossmaticeratidae and Scaphitidae). *Izv. AN. Armen. SSR Nauk o Zemle*, **5**, 31–42, 1 pl. [In Armenian.]
—— and KHAKHIMOV, F. K. 1976. Campanian and Maastrichtian ammonites in Central Asia. *Trudy Inst. geol. Dushanbe*, 146 pp., 13 pls. [In Armenian.]
BAILY, W. H. 1855. Description of some Cretaceous fossils from South Africa. Collected by Capt. Garden of the 45th Regiment. *Q. Jl geol. Soc. Lond.* **11,** 454–465, pls. 11–13.
BARRABÉ, L. 1929. Contribution à l'étude stratigraphique et pétrographique de la partie Médiane du pays de Sakalava, Madagascar. *Mém. Soc. géol. Fr.* N.S. **5,** Mém. 12, 269 pp., pls. 15–24 (1–10), map.
BASSE, E. 1931. Monographie paléontologique de Crétacé de la Province de Maintirano. *Mém. géol. Serv. Mines, Madagascar*, **1931,** 86 pp., 13 pls.
—— 1952. Ammonoides. *In* PIVETEAU, J. (ed.). *Traité de Paléontologie*, **2,** pp. 522–555, 581–688. Masson, Paris.
BAYLE, E. 1878. Fossiles principaux des terrains. *Explication de la Carte Géologique de France*, **4** (1), (Atlas), 158 pls. Service de la Carte Géologique détailée, Paris.
BESAIRIE, H. 1930. Recherches géologiques à Madagascar. Contribution à l'étude des ressources minerales. *Bull. Soc. Hist. nat. Toulouse*, **60,** 345–616, pls. 1–27.
BESKOVSKI, S. 1977. Sur la classification de la famille Desmoceratidae Zittel, 1895 (Ammonoidea, Crétacé). *Cr. Acad. bulg. Sci.* **30,** 891–894.
BILOTTE, M. 1984. Le Crétacé Supérieur des plate-formes Est-Pyrénéennes. *Strata*, **1,** 45 pls.

BILOTTE, M. 1985. Biozonation des séries de plate-form du Sénonien et du Maastrichtien est-pyrénéen. *Géol. Méditerran.* **10,** 99–102.

——and COLLIGNON, M. 1983. Biostratigraphie et paléontologie des ammonites du Sénonien Inférieur de Rennes-les-Bains-Sougraine (Aude) Zone sous-Pyrénéenne Orientale. *Docum. Lab. Géol. Lyon* H.S. **6,** 175–223, 7 pls.

BINKHORST, J. T. 1861. *Monographie des gastropodes et des céphalopodes de la Craie Supérieure du Limbourg,* vi+83 pp. (gastropods)+44 pp. (cephalopods), 17 pls. (1–5, 5a, $5a^1$, $5a^2$, $5a^3$, 5b, 5c, 5d, 6, 7, 8, 8a, 9). Muquardt, Brussels; Muller Frères, Maastricht.

BIRKELUND, T. 1965. Ammonites from the Upper Cretaceous of West Greenland. *Meddr Grønland,* **179,** 192 pp., 42 pls.

——HANCOCK, J. M., HART, M. B., RAWSON, P. F., REMANE, J., ROBASZYNSKI, F., SCHMID, F. and SURLYK, F. 1984. Cretaceous stage boundaries—proposals. *Bull. geol. Soc. Denmark,* **33,** 3–20.

BŁASZKIEWICZ, A. 1980. Campanian and Maastrichtian ammonites of the Middle Vistula Valley, Poland: a stratigraphic–paleontologic study. *Pr. Inst. geol.* **92,** 1–63, 56 pls.

BÖHM, J. 1898. Über *Ammonites pedernalis* V. BUCH. *Z. dt. geol. Ges.* **1898,** 183–201, pls. 5–7.

BOULE, M., LEMOINE, P. and THÉVENIN, A. 1906–1907. Paléontologie de Madagascar III. Céphalopodes crétacés des environs de Diego-Suarez. *Annls Paléont.* **1,** (1906), 173–192 (1–20), pls. 14–20 (1–7); **2** (1907), 1–56 (21–76), pls. 1–8 (8–15).

BREISTROFFER, M. 1947. Notes de nomenclature paléozoologique. *P.-v. mens. Soc. Sci. Dauphine,* **26** (195), 5 pp. [unpaginated.]

——1953. L'évolution des Turrilitidés Albiens et Cénomaniens. *C.r. hebd. Séanc. Acad. Sci.* (D) **234,** 2633–2635.

BRINKMANN, R. 1935. Die Ammoniten der Gosau und des Flysch in den nördlichen Ostalpen. *Mitt. geol. StInst. Hamb.* **15,** 1–14.

BRUNNSCHWEILER, R. O. 1966. Upper Cretaceous ammonites from the Carnavon Basin of Western Australia. 1. The heteromorph Lytoceratina. *Bull. Bur. Miner. Resour. Geol. Geophys. Aust.* **58,** 58 pp., 8 pls.

CASEY, R. 1960. *Hengestites,* a new genus of Gault ammonites. *Palaeontology,* **2,** 200–209, pl. 29.

——1965. A monograph of the Ammonoidea of the Lower Greensand. *Palaeontogr. Soc.* [*Monogr.*], Part 1, 399–546, pls. 69–70.

CHAVAN, A. 1947. La faune campanienne de Mont des Oliviers d'après les matériaux Vignal-Masse. *J. Conch. Paris,* **87,** 125–197, pls. 2–4.

CHIPLONKAR, G. W. and GHARE, M. A. 1976. Palaeontology of the Bagh Beds—Part VII: Ammonoidea. *Bull. Earth Sci.* for 1976, **4** and **5,** 1–10, 2 pls.

————1977. Palaeontology of the Bagh beds: part X—Scaphitidae. *Proc. Indian Acad. Sci.* **85B,** 67–76, 2 pls.

————1978. Taxonomic comments on Placenticeratidae with diagnosis of a new genus *Sancarlosia. Biovigyanam,* **4,** 75–80.

CHOFFAT, P. 1898. Recueil d'études paléontologiques sur la faune crétacique du Portugal. I, espèces nouvelles ou peu connues. Deuxième série, Les Ammonées du Bellasien, des couches à Neolobites Vibrayeanus, du Turonien et du Sénonien. *Trav. géol. Portugal,* (1898), 41–86, pls. (céphalopodes) 3–22.

COBBAN, W. A. 1964. The late Cretaceous cephalopod *Haresiceras* Reeside and its possible origin. *Prof. Pap. US geol. Surv.* **454-I,** 21 pp., 3 pls.

——1969. The late Cretaceous Ammonites *Scaphites leei* Reeside and *Scaphites hippocrepis* (De Kay) in the Western Interior of the United States. Ibid. **619,** 1–27, 5 pls.

——1973*a*. Occurrence of the late Cretaceous ammonite *Hoplitoplacenticeras* in Wyoming. Ibid. **475-C,** 60–62.

——1973*b*. The late Cretaceous ammonite *Trachyscaphites pulcherrimus* (Roemer) in New Jersey and Texas. *J. Res. US geol. Surv.* **1,** 695–700.

——1974. Ammonites from the Navesink Formation at Atlantic Highlands, New Jersey. *Prof. Pap. US geol. Surv.* **845,** 20 pp., 11 pls.

——and SCOTT, G. R. 1964. Multinodose scaphitid cephalopods from the lower part of the Pierre Shale and equivalent rocks in the conterminous United States. Ibid. **483-E,** 13 pp., 4 pls.

COLLIGNON, M. 1931. Faunes Senoniennes du Nord et de l'Ouest de Madagascar. *Annls. géol. Serv. Mines, Madagascar,* **1,** 7–64, pls. 1–9.

——1932. Paléontologie de Madagascar, XVII. Fossiles du Crétacé Supérieur du Menabe. *Annls Paléont.* **21,** 35–87 (1–53), pls. 4–12 (1–9).

——1938. Ammonites Campaniennes et Maastrichtiens de l'ouest et du sud de Madagascar. *Annls géol. Serv. Mines, Madagascar,* **9,** 55–118 (1–65), pls. 1–9.

——1948. Ammonites néocrétacées du Menabe (Madagascar). I. Les Texanitidae. Ibid. **13,** 49–107 (1–63), pls. 7–20 (1–14); **14**; 78–101 (64–120), pls. 15–23.

—— 1952. Ammonites néocrétacées du Menabe (Madagascar). II—Les Pachydiscidae. *Trav. Bur. géol. Madagascar*, **41**, 114 pp., 33 pls.

—— 1955. Ammonites néocrétacées du Menabe (Madagascar). II.—Les Pachydiscidae. *Annls géol. Serv. Mines, Madagascar*, **21**, 98 pp., 28 pls.

—— 1956. Ammonites néocrétacées du Menabe (Madagascar). IV—Les Phylloceratidae. V.—Les Gaudryceratidae. VI.—Les Tetragonitidae. Ibid. **23**, 1–106, pls. 1–11.

—— 1961. Ammonites néocrétacées du Menabe (Madagascar). VII—Les Desmoceratidae. Ibid. **31**, 115 pp., 32 pls.

—— 1966. *Atlas des fossiles caractéristiques de Madagascar (Ammonites).* **XIV** *(Santonien)*, x + 134 pp., pls. 455–513. Service Géologique, Tananarive.

—— 1969. *Atlas des fossiles caractéristiques de Madagascar (Ammonites).* **XV** *(Campanien inférieur)*, xi + 216 pp., pls. 514–606. Service Géologique, Tananarive.

—— 1970. *Atlas des fossiles caractéristiques de Madagascar (Ammonites).* **XVI** *(Campanien moyen; Campanien supérieur)*, iv + 82 pp., pls. 607–639. Service Géologique, Tananarive.

—— 1971. *Atlas des fossiles caractéristiques de Madagascar (Ammonites)* **XVII** *(Maestrichtien)*, iv + 44 pp., pls. 640–658. Service Géologique, Tananarive.

—— 1983. *In* BILOTTE, M. and COLLIGNON, M. 1983. q.v.

COQUAND, H. 1853. Déscription d'espèces nouvelles de coquilles fossiles du Musée de Besançon. *J. Conch. Paris*, **4**, 439–443, pl. 14.

—— 1856. Notice sur la formation crétacée du départment de la Charente. *Bull. Soc. géol. Fr.* (2) **14**, 55–98.

—— 1857*a*. Sur la craie supérieur de l'Aquitaine. Ibid. 743–745.

—— 1857*b*. Position des *Ostrea columba* et *biauriculata* dans le groupe de la craie inférieure. Ibid. 745–766.

—— 1857*c*. Réunion extraordinaire à Angoulême. Ibid. 841–903.

—— 1858. *Déscription physique, géologique, paléontologique et minéralogique du départment de la Charente* (imprimé sous la direction du Conseil Générale), **I**, 542 pp., Map. De Dodivers et Cie, Besançon.

—— 1859. Synopsis des animaux et des végétaux fossiles observés dans la formation crétacée du Sud-Ouest de la France. *Bull. Soc. géol. Fr.* (2) **16**, 945–1023.

—— 1860. *Description physique, géologique, paléontologique et minéralogique du Départment de la Charente* (imprimé sous les auspices du Conseil Générale), **II**, 420 pp. Barlatier-Feissat et Demonchy, Marseille.

—— 1861. Rapports qui existent entre les groupes de la craie moyenne et de la craie supérieur de la Provence et du Sud-Ouest de la France. *Bull. Soc. géol. Fr.* (2) **18**, 133–162.

DEKAY, J. E. 1828. Report on several fossil multilocular shells from the state of Delaware: with observations on a second specimen of the new fossil genus EURYPTERUS. *Ann. Lyceum nat. Hist.* **2**, 273–279, pl. 5, figs. 2–5 [only].

DESMAREST, A. G. 1817. Mémoire sur deux genres de coquilles fossiles cloisonnées et à siphon. *J. Phys. Chim. Hist. nat.* **85**, 42–51, pls. 1 and 2.

DIENER, C. 1925. Ammonoidea Neocretacea. *Fossilium Cat.* (1: Animalia), **29**, 244 pp.

DIETRICH, W. O. 1938. Zur Stratigraphie der Kreide im nördlich Zululand. *Zentbl. Miner. Geol. Paläont.* (B) **1938**, 228–240.

DOUVILLÉ, H. 1890. Sur la classification des cératites de la Craie. *Bull. Soc. géol. Fr.* (3) **18**, 275–292.

—— 1904. *Mission scientifique en Perse par J. de Morgan.* **3**—Études géologiques Partie IV.—Paléontologie Deuxième Partie, Mollusques fossiles, 191–380, pls. 25–50. Leroux, Paris.

DUJARDIN, F. 1837. Mémoire sur les couches du sol en Touraine et déscription des coquilles de la craie et des Faluns. *Mém. Soc. géol. Fr.* **2**, 211–311, pls. 15–20.

DUMONT, A. H. 1849. Rapport sur la carte géologique du Royaume. *Bull. Acad. r. Belg. Cl. Sci.* **16**, 351–373.

ERNST, G. and SCHMID, F. 1979. Multistratigraphische Untersuchungen in der Oberkreide des Raumes Braunschweig-Hannover. *Aspekte der Kreide Europas* IUGS Series **A6**, 11–46.

FAVRE, E. 1869. *Déscription des mollusques fossiles de la Craie des environs de Lemberg en Galicie*, xii + 187 pp., 13 pls. H. Georg, Geneva.

FIEBER, I. X. 1853. Synopsis der Europäischen Orthopteren mit besonderes Rücksicht der Böhmischen Arten. *Lotos*, **3**, 90–104, 115–129, 138–154, 168–176, 184–188, 201–207, 232–238, 252–261.

FORBES, E. 1846. Report on the fossil Invertebrata from Southern India, collected by Mr. Kaye and Mr. Cunliffe. *Trans. geol. Soc. Lond.* (2) **7**, 97–174, pls. 7–19.

FRECH, F. 1915. Uber Scaphites. 1. Die Bedeutung von *Scaphites* für die Gliederung der Oberkreide. *Zentbl Miner. Geol. Paläont.* **1915**, 553–568.

FRITSCH, A. and KAFKA, J. 1887. *Die Crustaceen der böhmischen Kreideformationen*, 53 pp., 10 pls. Fr. Řivnač, Prague.

GALE, A. S. 1980. Penecontemporaneous folding, sedimentation and erosion in Campanian Chalk near Portsmouth, England. *Sedimentology*, **27**, 137–151.

GEINITZ, H. B. 1843. *Die Versteinerungen von Kieslingwalde und Nachtrag zur Charakteristik des Sächsisch-böhmischen Kreidegebirges*, 23 pp., 6 pls. der Arnoldischen Buchhandlung, Dresden and Leipzig.

—— 1846. *Grundriss der Versteinerungskunde*, 813 pp., 26 pls. der Arnoldischen Buchhandlung, Dresden and Leipzig.

—— 1849–1850. *Das Quadersandsteingebirge oder Kreidegebirge in Deutschland*, 293 pp., 12 pls. Craz and Gerlach, Freiberg.

—— 1871–1875. Das Elbthalgebirge in Sachsen. *Palaeontographica*, **A20,** 1–319, pls. 1–67 (1–94, pls. 1–23, 1871; 95–207, pls. 24–45, 1872; 207–236, pls. 46–52, 1873; 237–276, pls. 53–60, 1874; 277–319, pls. 61–67, 1875).

GERTH, H. 1961. Neue Ammoniten funde in den Gossauschichten der Gosau und ihre stratigraphische Bedeutung. *Neues Jb. Geol. Paläont. Abh.* **112,** 119–142, pl. 24.

GIERS, R. 1964. Die Grossfauna der Mukronatenkreide (unteres Obercampan) im östlichen Münsterland. *Fortschr. Geol. Rheinld. Westf.* **7,** 213–294, 8 pls.

GILL, T. 1871. Arrangement of the Families of Mollusks. *Smithson. misc. Collns*, **227,** xvi+49 pp.

GILLARD, P. A. 1943. Sur la présence de Bélemnitelles (*B. mucronata* Schloth.) dan l'Aturien supérieur de la Charente. *C.r. somm. Séanc. Soc. géol. Fr.* (5) **13,** 157–159.

GRIEPENKERL, O. 1889. Die Versteinerungen der Senonen Kreide von Königslutter im Herzogtum Braunschweig. *Palaeont. Abh.* **4,** 305–419 (3–116), pls. 34–45 (1–12).

GROSSOUVRE, A. DE. 1894. Recherches sur la craie supérieure, 2, Paléontologie. Les ammonites de la craie supérieure. *Mém. Serv. Carte géol. dét. Fr.* 264 pp., 39 pls. [misdated 1893].

—— 1901. Recherches sur la craie supérieure 1: stratigraphie générale. Ibid. vii + 1,013 pp.

—— 1908*a*. Déscription des ammonites du Crétacé Supérieur du Limbourg Belge et Hollandais et du Hainault. *Mém. Mus. r. Hist. nat. Belg.* **4,** 1–39, pls. 1–11.

—— 1908*b*. Sur l'age des calcaires de Contes-les-Pins et de la zone à *Placenticeras bidorsatum* et *Mortoniceras delawarense*. *Bull. Soc. géol. Fr.* (4) **8,** 311.

—— 1930. Note sur le Bathonien moyen. *Livre Jubilaire, Centenaire Soc. géol. Fr.* **2,** 361–387, pls. 39–40.

HALLAM, A., HANCOCK, J. M., LA BRECQUE, J. L., LOWRIE, W. and CHANNELL, J. E. T. 1985. Jurassic to Palaeogene: Part 1 Jurassic and Cretaceous geochronology and Jurassic to Palaeogene magnetostratigraphy, 118–140. *In* SNELLING, N. J. (ed.). *The Chronology of the Geological Record. Geol. Soc. Lond., Memoir* **10,** 343 pp.

HANCOCK, J. M. and KENNEDY, W. J. 1981. Upper Cretaceous ammonite stratigraphy: some current problems. *In* HOUSE, M. R. and SENIOR, J. R. (eds.). *The Ammonoidea. Spec. Vol. Syst. Ass.* **18,** 531–553.

HASSAN, M. Y. 1971. New molluscan fauna from the Maastrichtian of Kharga Oasis, South Western Desert of Egypt. *Proc. Egypt. Acad. Sci.* **23,** 65–75, pls. 1 and 2.

HAUER, F. VON. 1858. Ueber die Cephalopoden der Gosauschichten. *Beitr. Paläont. Öest.* **1,** 7–14, pls. 2–4.

—— 1866. Neue Cephalopoden aus den Gosaugebilden der Alpen. *Sbor. Akad. Wiss. Wien. math. nat. Cl.* **53,** 1–9, pls. 1 and 2.

HAUG, E. 1907–1911. *Traité de Géologie*, 2,024 pp., 135 pls. Masson, Paris.

HENDERSON, R. A. 1970. Ammonoidea from the Mata Series (Santonian–Maastrichtian) of New Zealand. *Spec. Pap. Palaeont.* **6,** 82 pp., 15 pls.

—— and MCNAMARA, K. J. 1985. Maastrichtian non-heteromorph ammonites from the Miria Formation, western Australia. *Palaeontology*, **28,** 35–88, pls. 1–9.

HERM, D., KAUFFMAN, E. G. and WIEDMANN, J. 1979. The age and depositional environment of the 'Gosau'-Group (Coniacian–Santonian), Brandenberg/Tirol, Austria. *Mitt. Bayer. St. Paläont. Hist. Geol.* **19,** 27–92, pls. 1–11.

HINTE, J. E. VAN. 1979. The Coniacian, Santonian, and Campanian stratotypes. *Lethaia*, **12,** 183–187.

HOEPEN, E. C. N. VAN. 1920. Description of some Cretaceous ammonites from Pondoland. *Ann. Transv. Mus.* **7,** 142–147, pls. 24–26.

—— 1968. New Ammonites from Zululand. *Ann. geol. Surv. S. Afr.* **4,** 183–186, 7 pls. (1965).

HOLZAPFEL, E. 1887–1888. Die Mollusken der Aachener Kreide. *Palaeontographica*, **A4,** 29–180, pls. 4–21.

HOURCQ, V. 1950. Paléontologie de Madagascar XXVIII. Sur quelques ammonites du Sénonien. *Annls Paléont.* **35,** 89–117 (3–31), pls. 11–13 (1–3).

HOUŠA, V. 1967. *Lewesiceras* Spath (Pachydiscidae, Ammonoidea) from the Turonian of Bohemia. *Sb. geol. Věd. Praha*, **P9,** 7–49, 8 pls.

HOWARTH, M. K. 1965. Cretaceous ammonites and nautiloids from Angola. *Bull. Br. Mus. nat. Hist.* (Geol.), **10,** 335–412, 13 pls.

HUPSCH, J. W. C. A. F. 1768. *Neue in der Naturgeschichte des Niederdeutschlands gemachte Entdeckungen einiger selten und Wenig bekanten versteinerten Schaalthiere*, 159 pp., 9 pls. der Metternichisschen Buchhandlung, Frankfurt and Leipzig.

HYATT, A. 1889. Genesis of the Arietidae. *Smithson. Contr. Knowl.* **673,** xi + 238 pp., 14 pls.

—— 1894. Phylogeny of an acquired characteristic. *Proc. Am. phil. Soc.* **32,** 349–647, pls. 1–14.

—— 1900. Cephalopoda, pp. 502–604. *In* ZITTEL, K. A. VON. 1896–1900, *Textbook of Palaeontology*, transl. EASTMAN, C. R. Macmillan, London and New York.

—— 1903. Pseudoceratites of the Cretaceous. *Monogr. US geol. Surv.* **44,** 351 pp., 47 pls.

IMKELLER, H. 1901. Die Kreidebildungen und ihre Fauna am Stallauer Eck und Enzenauer Kopf bei Tölz. *Palaeontographica*, **A48,** 1–64, 3 pls.

IMMEL, H., KLINGER, H. C. and WIEDMANN, J. 1982. Die Cephalopoden des Unteren Santon der Gosau von Brandenberg/Tirol, Österreich. *Zitteliana*, **8,** 3–32, 11 pls.

INTERNATIONAL COMMISSION ON ZOOLOGICAL NOMENCLATURE. 1964. *International Code of Zoological Nomenclature adopted by the XV International Congress of Zoology*, 2nd ed., xix + 176 pp. International Trust for Zoological Nomenclature, London.

—— 1985. Ibid. 3rd edn., xx + 338 pp. International Trust for Zoological Nomenclature, London.

JIMBO, K. 1894. Beiträge zur Kenntniss der fauna der Kreideformation von Hokkaido. *Paläont. Abh.* (N.S.) **2,** 147–194, pls. 17–25.

JONES, D. L. 1963. Upper Cretaceous (Campanian and Maestrichtian) ammonites from southern Alaska. *Prof. Pap. US geol. Surv.* **432,** 53 pp., 40 pls.

KAYSER, E. 1924. *Lehrbuch der Geologie* **II,** *Geologischen Formationskunde*, 7th edn., viii + 675 pp. Ferdinand Enke, Stuttgart.

KENNEDY, W. J. 1984*a*. Systematic palaeontology and stratigraphic distribution of the ammonite faunas of the French Coniacian. *Spec. Pap. Palaeont.* **31,** 160 pp., 33 pls.

—— 1984*b*. Ammonite faunas and the 'standard zones' of the Cenomanian to Maastrichtian Stages in their type areas, with some proposals for the definition of stage boundaries by ammonites. *Bull. geol. Soc. Denmark*, **33,** 147–161.

—— 1985. Ammonite faunas of the Coniacian, Santonian and Campanian stages in the Aquitaine Basin. *Géol. Méditerran.* **10,** 103–113.

—— 1986*a*. The ammonite fauna of the Calcaire à *Baculites* (Upper Maastrichtian) of the Cotentin Peninsula (Manche, France). *Palaeontology*, **29,** 25–83, pls. 1–16.

—— 1986*b*. The Campanian–Maastrichtian ammonite sequence in the environs of Maastricht (Limburg, the Netherlands), Limburg and Liège Provinces (Belgium). *Newsl. Stratigr.* **16,** 149–168.

—— 1986*c*. In KENNEDY, W. J., BILOTTE, M., LEPICARD, B. and SEGURA, F. 1986, q.v.

—— in press *a*. Ammonites from the type Santonian and adjacent parts of northern Aquitaine (western France). *Palaeontology*.

—— in press *b*. The ammonite fauna of the type Maastrichtian with a revision of *Ammonites colligatus* Binkhorst, 1861. *Bull. Inst. r. Sci. nat. Belg.*

—— BILOTTE, M., LEPICARD, B. and SEGURA, F. 1986. Upper Campanian and Maastrichtian Ammonites from the Petites Pyrénées. *Eclog. geol. Helv.* **79,** 1001–1037, 5 pls.

—— and COBBAN, W. A. 1976. Aspects of ammonite biology, biogeography and biostratigraphy. *Spec. Pap. Palaeont.* **17,** 94 pp., 11 pls.

—— KAUFFMAN, E. G. and KLINGER, H. C. 1974. Upper Cretaceous invertebrate faunas from Durban, South Africa. *Trans. geol. Soc. S. Afr.* **77,** 95–111, 6 pls.

—— and KLINGER, H. C. 1977. Cretaceous faunas from Zululand and Natal, South Africa. The ammonite family Tetragonitidae Hyatt, 1900. *Ann. S. Afr. Mus.* **73,** 149–197.

—— and SUMMESBERGER, H. 1986. Lower Maastrichtian ammonites from Neuberg, Steirmark, Austria. *Beitr. Paläont. Östr.* **12,** 181–242, 16 pls.

—— and WRIGHT, C. W. 1983. *Ammonites polyopsis* Dujardin, 1837 and the Cretaceous ammonite family Placenticeratidae Hyatt, 1900. *Palaeontology*, **26,** 855–873, pls. 85–87.

KHAKHIMOV, F. K. 1972. Campanian Scaphitidae from western Tadzhikistan. *In* Lithology and stratigraphy of the oil and gas bearing strata of the Tadzhikistan basin. Dushanbe. [In Armenian.] (not seen)

KILIAN, W. and REBOUL, P. 1909. Les céphalopodes néocrétacés des îles Seymour et Snow Hill. *Wiss. Ergenbn. schwed. Südpolarexped.* **3** (6), 1–75, pls. 1–20.

KLINGER, H. C. 1976. Cretaceous heteromorph ammonites from Zululand. *Mem. geol. Surv. Rep. S. Afr.* **69,** 142 pp., 43 pls.

—— 1982. Revision of *Ancyloceras bipunctatum* Schlüter, 1872 (Cephalopoda, Ammonoidea) and discussion of the validity, phylogeny and limits of the genus *Neancyloceras* Spath, 1926. *Ann. S. Afr. Mus.* **90,** 219–239.

—— and KENNEDY, W. J. 1977. Upper Cretaceous ammonites from a borehole near Richards Bay, South Africa. Ibid. **72,** 69–107.

KLINGER, H. C., and KENNEDY, W. J. 1980. Cretaceous faunas from Zululand and Natal, South Africa. The ammonite family Texanitinae Collignon, 1948. *Ann. S. Afr. Mus.* **80,** 1–357.

KNER, R. 1848. Versteinerungen des Kreidemergels von Lemburg und seiner Umgebung. *Haidingers naturwiss. Abh.* **2,** 1–42, 5 pls.

KOSSMAT, F. 1895–1898. Untersuchungen über die Südindische Kreideformation. *Beitr. Paläont. Geol. Ost.-Ung.* **9,** (1895): 97–203 (1–107), pls. 15–25 (1–11); **11,** (1897); 1–46 (108–153), pls. 1–8 (12–19); **11,** (1898): 89–152 (154–217), pls. 14–19 (20–25).

KULLMANN, J. and WIEDMANN, J. 1970. Significance of sutures in phylogeny of Ammonoidea. *Paleont. Contr. Univ. Kans.* **44,** 1–32.

LAMARCK, J. P. B. A. DE M. DE. 1799. Prodrome d'une nouvelle classification des coquilles. *Mém. Soc. Hist. nat. Paris*, for 1799, 63–90.

—— 1801. *Système des Animaux sans vertebrès*, vii + 432 pp. The author; Deterville, Paris.

—— 1822. *Histoire naturelle des Animaux sans vertebrès*, **7,** 711 pp. Verdière, Paris.

LAMBERT, B. 1980. Étude de la nannoflore calcaire du Campanien charentais. *Cah. Micropal.* **3,** 39–53.

—— 1981. *Étude systematique des nannofossiles calcaires du Crétacé supérieur stratotypique (Coniacien, Santonien, Campanien). Répartition stratigraphique et biozonations.* Thèse Doct. État., Univ. Paris, 191 pp., 32 pls.

LAPEYRE, J. F. and THOMEL, G. 1972. *Scaphites genyi* nov. sp. du Campanien des environs de Contes (Alpes-Maritimes). *Annls Mus. Hist. nat. Nice*, **1,** 45–48, 2 pls.

LEANZA, A. F. 1964. *Parabinneyites*, nuevo nombre generico para *Patagoniceras* Leanza, 1963 non Wetzel, 1960. *Ameghiniana*, **19,** 84.

LEWY, Z. 1969. Late Campanian heteromorph ammonites from southern Israel. *Israel J. Earth Sci.* **18,** 109–135, pls. 1–4.

LINK, H. F. 1807. *Beschreibung der Naturalien-Sammlung der Universitätt zu Rostock. Vierte Abteilung: Fossilie Ueberbleibsel organischer Körper, sogennate Versteinerungen*, 30 pp., Rostock.

LOPUSKI, C. 1911. Przycznki do znajomosci fauny kredowej gubernii Lubelskiej. *C.r. Séanc. Soc. Sci. Varsovie*, **4,** 104–140, pls. 1–4.

MANTELL, G. A. 1822. *The fossils of the South Downs; or illustrations of the geology of Sussex*, vi + 327 pp., 42 pls. Lupton Relfe, London.

MARCINOWSKI, R. 1972. Condensed Cretaceous sequence with stromatolites in the Polish Jura chain. *Acta geol. pol.* **22,** 515–539, 4 pls.

—— 1980. Cenomanian ammonites from German Democratic Republic, Poland, and the Soviet Union. Ibid. **30,** 215–325, 20 pls.

MARIANI, E. 1898. Ammoniti del Senoniano lombardo. *Memorie Inst. lomb. Sci. Lett.* **18,** 51–61, pl. 8.

MARSHALL, P. 1926. The Upper Cretaceous ammonites of New Zealand. *Trans. NZ Inst.* **56,** 129–210, pls. 19–47.

MARTINEZ, R. 1982. Ammonoideos cretacicos del Prepirineo de la Provincia de Lleida. *Publicaciones de Geologia, Universidad Autonoma de Barcelona*, **17,** 197 pp., 30 pls.

MATSUMOTO, T. 1938. A biostratigraphic study on the Cretaceous deposits of the Naibuchi Valley, South Karahuto. *Proc. imp. Acad. Japan*, **14,** 190–194.

—— 1947. A note on the Japanese Pachydiscinae. *Sci. Rep. Dept. geol. Fac. Sci. Kyushu Univ.* **2,** 34–46. [In Japanese.]

—— 1951. A note on the Pachydiscinae, a Cretaceous ammonite-group. *Trans. Proc. palaeont. Soc. Japan*, N.S., **1,** 19–26.

—— 1954. Selected leading Cretaceous ammonites in Hokkaido and Saghalien, 243–313, pls. 17–36 (1–20). *In* MATSUMOTO, T. 1954. (ed.). *The Cretaceous System in the Japanese Islands*, 324 pp., 36 pls. Japan Society for the Promotion of Science, Tokyo.

—— 1955*a*. Family Kossmaticeratidae from Hokkaido and Saghalien. *Jap. J. Geol. Geogr.* **26,** 115–164, pls. 8–10.

—— 1955*b*. The bituberculate pachydiscids from Hokkaido and Saghalien. *Mem. Fac. Sci. Kyushu Univ. Ser. D. Geol.* **5,** 153–184, pls. 31–37.

—— 1956. Further notes on Kossmaticeratids from Hokkaido. *Jap. J. Geol. Geogr.* **27,** 173–187, pls. 14–16.

—— 1959*a*. Upper Cretaceous ammonites of California. Part II. *Mem. Fac. Sci. Kyushu Univ. Ser. D. Geol.* Special Volume, **1,** 172 pp., 41 pls.

—— 1959*b*. The Upper Cretaceous ammonites of California. Part 1. Ibid. **8,** 91–171, pls. 30–45.

—— 1967. Evolution of Nostoceratidae. Ibid. **18,** 331–347, pls. 18–19.

—— 1979. Some new species of *Pachydiscus* from the Tobetsu and the Hobetsu Valleys, 50–64. *In* MATSUMOTO, T., KANIE, Y. and YOSHIDA, S. Notes on *Pachydiscus* from Hokkaido. Ibid. **24,** 47–73, pls. 8–13.

—— 1982. Note on *Hoplitoplacenticeras* from Hokkaido. *Proc. Japan Acad.* **B58,** 249–252.

—— 1984*a*. A new tetragonitid ammonite from Hokkaido. Ibid. **B60,** 33–35.

—— 1984*b*. Some ammonites from the Campanian (Upper Cretaceous) of northern Hokkaido. *Spec. Pap. Palaeont. Soc. Japan*, **27,** 93 pp., 31 pls.

—— MURAMOTO, T. and INOMA, A. 1972. Two small desmoceratid ammonites from Hokkaido. *Trans. Proc. palaeont. Soc. Japan*, N.S., **87,** 377–394, pl. 47.

—— and OBATA, I. 1955. Some Upper Cretaceous desmoceratids from Hokkaido and Saghalien. *Mem. Fac. Sci. Kyushu Univ. Ser. D, Geol.* **5,** 119–151, pls. 24–30.

MEEK, F. B. 1862. Descriptions of new Cretaceous fossils collected by the Northwestern Boundary Commission on Vancouver and Sucia Islands. *Proc. Acad. nat. Sci. Philad.* **13,** 314–374.

—— 1871. A preliminary list of fossils collected by Dr. Hayden in Colorado, New Mexico and California, with brief descriptions of a few of the new species. *Proc. Am. Phil. Soc.* **11,** 425–431 (1869–1870).

—— 1876*a*. A report on the invertebrate Cretaceous and Tertiary fossils of the upper Missouri country. *In* HAYDEN, F. V. *Report of the United States Geological Survey of the Territories*, **9,** lxiv + 629 pp., 45 pls.

—— 1876*b*. Descriptions and illustrations of fossils from Vancouver and Sucia Islands and other north-western localities. *Bull. US geol. geogr. Surv. Territ.* **2** (4), 351–374, pls. 2–6.

MENEGHINI, G. 1854. Nuovi fossili Toscani. *Annali. Univ. tosc.* (2) **3,** 55–75.

MIKHAILOV, N. P. 1951. Upper Cretaceous ammonites from the southern part of European Russia and their importance for zonal stratigraphy (Campanian, Maastrichtian). *Trudy Inst. geol. Nauk. Akad. Nauk. SSSR*, **129** (Geol. Ser. 50), 143 pp., 19 pls. [In Russian.]

MOBERG, J. C. 1884–1885. Sveriges Kritsystem Systematisk Framställdt. *Sver. geol. Unders Afh C*, **63,** 45 pp., 2 pls. (1884); artbeskrifnig *ibid.* **73,** 64 + 2 pp., 6 pls. (1885).

MORTON, S. G. 1830. Synopsis of the organic remains of the Ferruginous Sand Formation of the United States, with geological remarks. *Am. J. Sci.* **17,** 274–295, 3 pls.

—— 1834. *Synopsis of the organic remains of the Cretaceous group of the United States. Illustrated by nineteen plates, to which is added an appendix containing a tabular view of the Tertiary fossils discovered in America*, 88 pp., 19 pls. Key and Biddle, Philadelphia.

—— 1842. Description of some new species of organic remains of the Cretaceous Group of the United States; with a tabular view of the fossils hitherto discovered in this formation. *J. Acad. nat. Sci. Philad.* **8,** 207–227.

MOURLON, M. 1878. *Mémoires sur les terrains Crétacé et Tertiaires préparés par feu* ANDRÉ DUMONT, *pour servir à la déscription de la carte géologique de la Belgique*, **I,** *Térrain Crétacé*, xx + 556 pp.; **II,** *Terrain Tertiaires*, 459 pp. Musée Royal d'Histoire Naturelle de Belgique, Brussels.

MÜLLER, G. and WOLLEMANN, A. 1906. Die Molluskenfauna des Untersenon von Braunschweig und Ilsede. II. Die Cephalopoden. *Abh. preuss. geol. Landesanst.* **47,** 1–30, pls. 1–11.

NAIDIN, D. P. 1974. Ammonoidea. *In* KRYMGOLTS, G. JA. (ed.). *Atlas of Upper Cretaceous fauna of Donbass*, 158–195, pls. 53–70. 'NEDRA', Moscow. [In Russian.]

—— and SHIMANSKIJ, V. N. 1959. Cephalopoda. *In* MOSKVINA, M. M. (ed.). *Atlas of the Upper Cretaceous fauna of the northern Caucasus and Crimea*, 166–220, 23 pls. Moscow. [In Russian.]

NALIVAJKO, L. E. 1936. *Makrofauna goryishn' okrejdovikh pokladyiv nyivdennoyi okrayini Donbassa*, **2,** 61 pp., 16 pls. Krayins'kij geologyichnij trest, Kiyiv. [In Ukrainian.]

NEUMANN, M. 1980*a*. Observations micropaléontologiques à propos du Campanien et du Maestrichtien. *Neues Jb. Geol. Paläont. Mh.* **7,** 417–427.

—— 1980*b*. Le Campanien. *Mém. Bur. Rech. géol. minièr*, **109,** 161–170.

—— 1980*c*. Répartition stratigraphique des grands Foraminifères du Campanien et du Maastrichtien en Mesogée. *Int. geol. Congr.* **1** (4), p. 266.

—— ANDREIEFF, P., LAMBERT, B. and PLATEL, J. P. 1984. Une exemple précis du passage Campanien–Maastrichtien en facies néritique: la région de Maurens, Dordogne (France). *C.r. hebd. Séanc. Acad. Sci., Paris*, **298,** 845–850.

—— and PLATEL, J. P. (Coordonnateurs) 1985*a*. Le Campanien stratotypique: étude lithologique et micro-paléontologique. *Géol. Méditerran.* **10,** 41–57.

—— —— 1985*b*. Synthèse bibliographique de la répartition des céphalopodes dans le Sénonien supérieur d'Aquitaine septentrionale. Ibid. **10,** 115–120.

NOWAK, K. 1908. Untersuchungen über die Cephalopoden der oberen Kreide in Polen. I. Teil. Genus *Baculites* Lamarck. *Bull. int. Acad. Sci. Lett. Cracovie, Cl. Sci. math. nat., B* for **1908,** 326–353, pl. 14.

—— 1911. Untersuchungen über die Cephalopoden der oberen Kreide in Polen. II. Teil. Die Skaphiten. Ibid. for **1911,** 547–589, pls. 32 and 33.

NOWAK, K. 1913. Untersuchungen über die Cephalopoden der oberen Kreide in Polen. III. Teil. *Bull. int. Acad. Sci. Lett. Cracovie, Cl. Sci. math. nat., B* for **1913,** 335–415, pls. 40–45.

—— 1916. Zur Bedeutung von *Scaphites* für die Gliederung der Oberkreide. *Verhandl. geol. Reichsanst.* (*St Anst-Landesanst*), *Wien* for **1916,** 55–67.

OBATA, I., FUTAKAMI, M., KAWASHITA, Y. and TAKAHASHI, T. 1978. Apertural features in some Cretaceous ammonites from Hokkaido. *Bull. natn. Sci. Mus. Tokyo*, **4,** 139–154, 3 pls.

ØDUM, H. 1953. De geologiska resultaten från borringarna vid Hollviken. *Sver. geol. Unders. Afh.* **C46,** no. 527, 37 pp., 4 pls.

OLSSEN, A. A. 1944. Contributions to the palaeontology of northern Peru. The Cretaceous of the Paita region. *Bull. Am. Paleont.* **28,** 1–164, pls. 1–17.

ORBIGNY, A. D'. 1840–1842. *Paléontologie française: Terrains crétacés.* **1.** *Céphalopodes.* 1–120 (1840); 121–430 (1841); 431–662 (1842), 148 + 3 pls. Masson, Paris.

—— 1850. *Prodrome de Paléontologie stratigraphique universelle des animaux mollusques et rayonnés*, **2,** 428 pp. Masson, Paris.

—— 1852. *Cours elementaire de paléontologie et de géologie stratigraphique*, **2,** 383–847. Masson, Paris.

OWEN, D. D. 1852. *Geological Survey of Wisconsin, Iowa and Minnesota, and incidentally a portion of Nebraska Territory: made under the direction of the U.S. Treasury Department Philadelphia*, 195 pp., 29 pls. + maps and sections. Lippincott, Grambo and Co., Philadelphia.

PARKINSON, J. 1811. *Organic remains of a former world*, **3,** xvi + 479 pp., 22 pls. Sherwood, Neily and Jones, London.

PAULCKE, W. 1907. Die Cephalopoden der oberen Kreide Südpatagoniens. *Ber. naturw. Ges. Freiburg i.B.* **15,** 167–248, pls. 10–19.

PERVINQUIÈRE, L. 1907. Études de paléontologie tunisienne. 1. Céphalopodes des terrains sécondaires. *Carte géol. Tunisie*, v + 438 pp., 27 pls. De Rudeval, Paris.

PHILIP, J., and BILOTTE, M. 1985. Les Rudistes du Sénonien de la France. Précisions stratigraphiques sur le Dordonien. *Géol. Méditerran.* **10,** 183–192, 1 pl.

PHILLIPS, J. 1829. *Illustrations of the geology of Yorkshire*, x + 192 pp., 9 + 14 pls. Thomas Wilson, York.

PLATEL, J. P. 1977. Le Campanien stratotypique dans le synclinal de Saintes (Charentes): lithostratigraphie, géomorphologie et biozonation. *Bull. Bur. Rech. géol. min. Paris*, (2), (1) **4,** 261–276.

PRUVOST, A. 1910. Les ammonites Sénoniennes du Nord de la France. *Annls Soc. géol. N.* **39,** 365–368.

QUAAS, A. 1902. Beitrag zur Kenntnis der Fauna der Obersten Kreidebildungen in der Libysche Wüste (Overwegi-Schichten und Blattertone). *Palaeontographica*, **A30,** 153–336, pls. 20–33.

REDTENBACHER, A. 1873. Die Cephalopodenfauna der Gosauschichten im den nordöstlichen Alpen. *Abh. K.-K. geol. Reichsanst.* **5,** 91–140, pls. 22–30.

REESIDE, J. B. 1927. The cephalopods of the Eagle Sandstone and related formations in the Western Interior of the United States. *Prof. Pap. US geol. Surv.* **151,** 87 pp., 45 pls.

—— 1928. The Scaphites, an Upper Cretaceous ammonite group. Ibid. **150B,** 21–36, pls. 9–11.

—— 1962. Cretaceous ammonites of New Jersey. *Bull. New Jers. Bur. geol. Topogr.* **61,** 113–137, pls. 68–75.

REYMENT, R. A. 1958. Neubeschreibung der Redtenbacher'schen Ammoniten-originale aus den Gosauschichten. *Stockh. Contr. Geol.* **3,** 31–49, 12 pls.

RICHARDS, H. G. 1953. *Record of the rocks—the geological story of eastern North America*, 413 pp. The Ronald Press Company, New York.

RIEDEL, L. 1931. Zur Stratigraphie und Faciesbildung im Oberemscher und Untersenon am Südrande des Beckens von Münster. *Jb. preuss. geol. Landesanst. BergAkad.* **51,** 605–713, pls. 72–79.

ROEMER, F. A. 1840–1841. *Die Versteinerungen des norddeutschen Kreidegebirges*, 145 pp., 16 pls. (1–48, pls. 1–7, 1840; 49–145, pls. 8–16, 1841). Hahn'schen Hofbuchhandlung, Hannover.

ROLLIER, L. 1922. Phylogenie des Ammonites. *Eclog. geol. Helv.* **17,** 358–360, pls. 20–22.

ROMAN, F. 1938. *Les ammonites jurassiques et crétacées. Essai de genera*, 554 pp., 53 pls. Masson, Paris.

ROMEIN, B. J. 1962. On the type locality of the Maastrichtian (Dumont, 1849), the upper boundary of that stage and on the transgression of a Maastrichtian s.l. in southern Limburg. *Meded. geol. Sticht.* N.S. **15,** 77–84.

SCHLÜTER, C. 1867. *Beitrag zur Kenntniss der jüngsten Ammoneen Norddeutschlands*, 36 pp., 6 pls. A. Henry, Bonn.

—— 1871–1876. Cephalopoden der oberen deutschen Kreide. *Palaeontographica*, **21,** 1–24, pls. 1–8 (1871); **21,** 25–120, pls. 9–35 (1872*a*); **24,** 1–144 (121–264) + x, pls. 36–55 (1876).

—— 1872*b*. *Verh. naturh. Ver. preuss. Rheinl.* **29,** Correspondenzblatt 1, 91–92.

SCHMID, F. and ERNST, G. 1975. Ammoniten aus dem Campan und lehrter westmulde und ihre stratigraphische Bedeutung. *Ber. naturhist. Ges. Hannover*, **119,** 315–359, pls. 1–4.

SCHOLZ, G. 1978. Beitrag zur Kenntnis des ostalpinen Alb. 1. Oberalb-ammoniten aus dem Kampenwaldvorland (Chiemgauer Alpen). *Mitt. bayer. St. Paläont. Hist. Geol.* **18,** 39–44, pl. 3.

SCHULZ, M. G. 1978. Zur Litho- und Biostratigraphie des Obercampan–Untermaastricht von Lagerdorf und Kronsmoor (SW-Holstein). *Newsl. Stratigr.* **7,** 73–89.

—— 1979. Morphometrisch-variationsstatistische Untersuchungen zur Phylogenie der belemniten-gattung *Belemnella* im Untermaastricht NW Europas. *Geol. Jb.* **A47,** 3–157, 12 pls.

—— ERNST, G., ERNST, H. and SCHMID, F. 1984. Coniacian to Maastrichtian stage boundaries in the standard section for the Upper Cretaceous white chalk of NW Germany (Langerdorf–Kronsmoor–Hemmoor): definitions and proposals. *Bull. geol. Soc. Denmark,* **33,** 203–215.

—— and SCHMID, F. 1983. Das Ober-Maastricht von Hemmoor (N-Deutschland): Faunenzonen-Gliederung und Korrelation mit dem Ober-Maastricht von Danemark und Limburg. *Newsl. Stratigr.* **13,** 21–39.

SCOTT, G. R. and COBBAN, W. A. 1965. Geologic and biostratigraphic map of the Pierre Shale between Jarre Creek and Loveland, Colorado. *US geol. Surv. Misc. Geol. Inv. Map,* **I-439.**

SEGURA, F. 1979. *Étude géologique dans la partie orientale des Petites Pyrénées-Zones Sous-pyrénéennes-(Ariège, Haute-Garonne),* 143 pp. Thèse du Troisième Siècle, Univ. Paul Sabatier, Toulouse.

SEMENOV, W. P. 1899. The fauna of the Cretaceous deposits of Mangychlak and some other localities in the Transcaspian province. *Trav. Soc. Imp. St. Petersbourg,* **28** (5), *Sect. Géol et Min.* 1–178, pls. 1–5. [In Russian.]

SÉRONIE-VIVIEN, M. 1956. *Étude de Crétacé supérieur de la vallée d l'Isle.* Diplôme Études Sup., Université Bordeaux (*fide* Séronie-Vivien).

—— 1960. Les localites types du Sénonien dans les environs de Cognac et de Barbézieux (Charente). *C.r. congrès Soc. Sav. Dijon 1959: Colloque sur le Crétacé supérieur français.* 579–589.

—— 1972. *Contribution à l'etude du Sénonen en Aquitaine septentrionale. Ses strctotypes: Coniacien, Santonien, Campanien,* 195 pp., 16 pls. Éditions du Centre National de la Recherche Scientifique, Paris.

SEUNES, J. 1890. Contributions à l'étude des céphalopodes du Crétacé Supérieur de France. 1. Ammonites du Calcaire à *Baculites* du Cotentin. *Mém. Soc. géol. Fr. Paléont.* **1,** Mém. 2, 1–7, pls. 2–3 (1–2).

—— 1891. Contributions à l'étude des cephalopodes du crétacé Supérieur de France. I Ammonites du Calcaire à *Baculites* du Cotentin (Suite). II Ammonites du Campanien de la région sous-pyrénéenne. Département de Landes. Ibid. **2,** Mém. 2, 8–22, pls. 12–15 (3–6).

SEYED-EMAMI, K., FORSTER, R. and MOJTEHEDI, A. 1984. Ammoniten aus dem mittleren Cenoman von Nordöst-Iran (Koppeh-Dagh). *Neues Jb. Geol. Paläont. Mh.* **1984,** 159–172.

SHARPE, D. 1853–1857. Description of the fossil remains of Mollusca found in the Chalk of England. I, Cephalopoda. *Palaeontogr. Soc.* [*Monogr.*], 68 pp., 27 pls. (1–26, pls. 1–10 (1853); 27–36, pls. 11–16 (1855); 37–68, pls. 17–27 (1857).)

SHIMIZU, S. 1935. The Upper Cretaceous cephalopods of Japan. Part 1. *J. Shanghai Sci. Inst.* sec. 2, **2,** 159–226.

SORNAY, J. 1958. *Turrilites* (*Bostrychoceras*) *archiacianus* d'Orbigny. *Palaeont. Univers.* N.S. **5,** 2 pp.

SOWERBY, J. 1812–1822. *The Mineral Conchology of Great Britain,* **1,** pls. 1–9 (1812), pls. 10–44 (1813), pls. 45–78 (1814), pls. 79–102 (1815); **2,** pls. 103–114 (1815), pls. 115–150 (1816), pls. 151–186 (1817), pls. 187–203 (1818); **3,** pls. 204–221 (1818), pls. 222–253 (1819), pls. 254–271 (1820), pls. 272–306 (1821); **4,** pls. 307–318 (1821), pls. 319–383 (1822). The author, London.

SOWERBY, J. DE C. 1823–1846. *The Mineral Conchology of Great Britain* (continued), **4,** pls. 384–407 (1823); **5,** pls. 408–443 (1823), pls. 444–485 (1824), pls. 486–503 (1825); **6,** pls. 504–544 (1826); pls. 545–580 (1827), pls. 581–597 (1828), pls. 598–609 (1829); **7,** pls. 610–618 (1840), pls. 619–623 (1841), pls. 624–628 (1843), pls. 629–643 (1844), pls. 644–648 (1846). The author, London.

SPATH, L. F. 1921. On Cretaceous Cephalopoda from Zululand. *Ann. S. Afr. Mus.* **12,** 217–321, pls. 19–26.

—— 1922*a*. On the Senonian ammonite fauna of Pondoland. *Trans. R. Soc. S. Afr.* **10,** 113–147, pls. 5–9.

—— 1922*b*. On Cretaceous Ammonoidea from Angola, collected by Professor J. W. Gregory, D.Sc., F.R.S. *Trans. R. Soc. Edinb.* **53,** 91–160, 4 pls.

—— 1923. A monograph of the Ammonoidea of the Gault. *Palaeontogr. Soc.* [*Monogr.*], Part 1, pp. 1–72, pls. 1–4.

—— 1925*a*. On Upper Albian Ammonoidea from Portuguese East Africa, with an appendix on Upper Cretaceous Ammonites from Maputoland. *Ann. Transv. Mus.* **11,** 197–200, pls. 28–37.

—— 1925*b*. On Senonian Ammonoidea from Jamaica. *Geol. Mag.* **62,** 28–32, pl. 1.

—— 1926. On new ammonites from the English Chalk. Ibid. **63,** 77–83, table.

—— 1939. Problems of ammonite nomenclature. 6. The genus *Pachydiscus* Zittel. Ibid. **74,** 277–281.

—— 1953. The Upper Cretaceous cephalopod fauna of Grahamland. *Scient. Rep. Br. Antarct. Surv.* **3,** 1–60, pls. 1–13.

STEPHENSON, L. W. 1941. The larger invertebrates of the Navarro Group of Texas (Exclusive of Corals and Crustaceans and exclusive of the fauna of the Escondido Formation). *Univ. Tex. Bull.* **4101,** 641 pp., 95 pls.

STOLICZKA, F. 1863–1866. The fossil cephalopoda of southern India. Ammonitidae with revision of the Nautilidae &c. *Mem. geol. Surv. India Palaeont. indica,* **3** (1), 41–56, pls. 26–31 (1863); (2–5), 57–106, pls. 32–54 (1864); (6–9), 107–154, pls. 55–80 (1865); (10–13), 155–216, pls. 81–94 (1866).

SUMMESBERGER, H. 1979. Eine obersantone Ammonitenfauna aus dem Becken von Gosau (Oberösterreich). *Ann. Naturhistor. Mus. Wien,* **82,** 109–176, 15 pls.

—— 1980. Neue Ammoniten aus der Sandkalkbank der Hochmoosschichten (Obersanton; Gosau, Österreichs). Ibid. **83,** 275–283.

—— 1985. Ammonite zonation of the Gosau Group (Upper Cretaceous, Austria). Ibid. **87A,** 145–166.

SURLYK, F. 1975. Die Brachiopoden der Hemmoorer Schreibekreide. *Paläont. Ges.* **45,** Hahresversammlung in Hannover, 1 p.

—— 1982. Brachiopods from the Campanian–Maastrichtian boundary sequence, Kronsmoor (NW Germany). *Geol. Jb.* **A61,** 259–277.

THIEDIG, F. and WIEDMANN, J. 1976. Ammoniten und Alter der höheren Kreide (Gosau) des Krappefeldes in Kärnen (Osterreich). *Mitt. geol. StInst. Hamb.* **45,** 9–27, pls. 1 and 2.

TOMLIN, J. R. U. B. 1930. Some preoccupied generic names—II. *Proc. malac. Soc. Lond.* **19,** 22–24.

TUOMEY, M. 1856. Description of some new fossils from the Cretaceous rocks of the southern states. *Proc. Acad. nat. Sci. Philad.* **7,** 162–172.

TZANKOV, C. V. 1964. Ammonites from the Maastrichtian near Kladorub village, Belogradchic region, north-west Bulgaria. *Trudove Vărkhu geol. Būlg.* **6,** 143–168, pls. 1–10. [In Bulgarian.]

—— 1982. *Les Fossiles de Bulgarie Va. Crétacé Supérieur.* 136 pp., 50 pls. Bulgarian Academy of Sciences, Sofia. [In Bulgarian with French Summary.]

USHER, J. L. 1952. Ammonite faunas of the Upper Cretaceous of Vancouver Island, British Columbia. *Bull. geol. Surv. Can.* **21,** 1–182, pls. 1–30.

VENZO, S. 1936. Cefalopodi del Cretaceo medio-superiore dello Zululand. *Palaeont. Ital.* **36,** 59–133, pls. 5–12.

WARD, P. D. 1976. Upper Cretaceous ammonites (Santonian–Campanian) from Orcas Island, Washington. *J. Paleont.* **50,** 454–461, pl. 1.

WEDEKIND, R. 1916. Über Lobus, Suturallobus und Inzision. *Zentbl. Miner. Geol Paläont.* for **1916,** 185–195.

WEGNER, T. 1905. Die Granulatenkreide des Westlichen Münsterlandes. *Z. dt. geol. Ges.* **57,** 112–232, pls. 7–10.

WEIJDEN, W. J. M. VAN DER, 1943. Die Macrofauna der Hervenschen Kreide. *Meded. geol. Sticht.* **C42,** 4–139, pls. 1–15.

WHITEAVES, J. F. 1903. On some additional fossils from the Vancouver Cretaceous, with a revised list of species therefrom. *Geol. Surv. Canada Mesozoic Fossils,* **1** (5), 309–409, pls. 40–51.

WHITFIELD, R. P. 1892. Gasteropoda and Cephalopoda of the Raritan Clays and Greensand Marls of New Jersey. *Monogr. US geol. Surv.* **18,** 402 pp., 50 pls.

WIEDMANN, J. 1960. Le Crétacé supérieur de l'Espagne et du Portugal et ses céphalopodes. *C.r. congrès Soc. Sav. Dijon, 1959: Colloque sur le Crétacé supérieur français,* 709–764, 8 pls.

—— 1962. Ammoniten aus der Vascogotischen Kreide (Nordspanian). 1, Phylloceratina, Lytoceratina. *Palaeontographica,* **A118,** 119–237, pls. 8–14.

—— 1966. Stammesgeschichte und System den posttriadischen Ammonoideen; ein Überblick. *Neues Jb. Geol. Paläont. Abh.* **125,** 49–79, pls. 1–2; **127,** 13–81, pls. 3–6.

—— 1978. Eine paläogeographisch interessante Ammonitenfaunula aus der alpinen Gosau. *Eclog. Geol. Helv.* **71,** 663–675, 2 pls.

—— 1979. Die Ammoniten der NW-deutschen, Regensberger und Ostalpinen Oberkreide im Vergleich mit den Oberkreidefaunen des westlichen Mediterrangebeites. *In* WIEDMANN, J. (ed.). *Aspekte der Kreide Europas, Int. Un. geol. Sci.* (A) **6,** 335–350.

WOLLEMANN, A. 1902. Die Fauna der Lüneburger Kreide. *Abh. preuss. geol. Landesanst.* N.F. **37,** 111 pp., 7 pls.

WOODS, H. 1906. The Cretaceous fauna of Pondoland. *Ann. S. Afr. Mus.* **4,** 275–350, pls. 3–44.

WRIGHT, C. W. 1952. A classification of the Cretaceous ammonites. *J. Paleont.* **26,** 213–222.

—— 1953. Notes on Cretaceous ammonites. I. Scaphitidae. *Ann. Mag. nat. Hist.* (12) **6,** 473–476.

—— 1955. Notes on Cretaceous ammonites. II. The phylogeny of the Desmocerataceae and the Hoplitaceae. Ibid. (12) **8,** 561–575.

—— 1957. *In* MOORE, R. C. (ed.). *Treatise on Invertebrate Paleontology. Part. L, Mollusca 4, Cephalopoda Ammonoidea,* xxii + 490 pp. Geological Society of America and University of Kansas Press, New York and Lawrence, Kansas.

——and KENNEDY, W. J. 1984. The Ammonoidea of the Lower Chalk. *Palaeontogr. Soc.* [*Monogr.*], Part 1, 1–126, pls. 1–40.
——and MATSUMOTO, T. 1954. Some doubtful Cretaceous ammonite genera from Japan and Saghalien. *Mem. Fac. Sci. Kyushu Univ. Ser. D.* (*Geol.*), **4,** 107–134, pls. 7 and 8.
——and WRIGHT, E. V. 1951. A survey of the fossil Cephalopoda of the Chalk of Great Britain. *Palaeontogr. Soc.* [*Monogr.*], 41 pp.
YABE, H. 1921. *In* YABE, H. and SHIMIZU, S. 1921. Notes on some Cretaceous ammonites from Japan and California. *Sci. Rep. Tohoku Imp. Univ.* (2) **5,** 53–59, pls. 8–9.
——and SHIMIZU, S. 1924. A new species of *Brahmaites* from the Upper Cretaceous of S. Saghalien, with some remarks on the genus *Brahmaites*. *Jap. J. Geol. Geogr.* **76,** 77–80, pl. 13.
————1926. A study on the genus '*Parapachydiscus*'. *Proc. imp. Acad. Japan*, **2,** 171–173.
YOUNG, K. 1963. Upper Cretaceous ammonites from the Gulf Coast of the United States. *Univ. Tex. Bull.* **6304,** ix + 373 pp., 82 pls.
ZABORSKI, P. M. P. 1982. Campanian and Maastrichtian sphenodiscid ammonites from southern Nigeria. *Bull. Br. Mus. nat. Hist.* (Geol.), **36,** 303–332.
ZHELEV, ST. T. 1934. Geology of the Pleven Region. II. Palaeontology. *Spls. bulg. geol. Dushanbe*, **6,** 166–204, pls. 3–7. [In Bulgarian.]
ZITTEL, K. A. 1884. *Handbüch der Palaeontologie* . . . Abt. 1, 2, (Lief 3), *Cephalopoda*, 329–522. R. Oldenbourg, Munich and Leipzig.
——1895. *Grundzüge der Palaeontologie* (*Palaeozoologie*). vii + 972 pp. R. Oldenbourg, Munich and Leipzig.

W. J. KENNEDY
Geological Collections
University Museum
Parks Road
Oxford OX1 3PW

Typescript received 14 June 1986
Revised typescript received 19 July 1986